NIOSH
Case Studies in

Ergonomics

Edited by
Shirley A. Ness

Government Institutes, Inc.
Rockville, Maryland

Government Institutes, Inc., 4 Research Place, Suite 200, Rockville, Maryland 20850

Copyright © 1996 by Government Institutes. All rights reserved.

00 99 98 97 96   5   4   3   2   1

These investigations and reports were conducted under the National Institute for Occupational Safety and Health, Centers for Disease Control of the U.S. Department of Health and Human Services under the authority of the Occupational Safety and Health Act of 1970. Government Institutes and the editor determined that these public domain reports were of interest to the general public and compiled them and added an introduction, detailed contents and index.

The editor and publisher make no representation of warranty, express or implied, as to the completeness, correctness or utility of the information in this publication. In addition, the editor and publisher assume no liability of any kind whatsoever resulting from the use of or reliance upon the contents of this book. All requests for permission to reproduce material from this work should be directed to Government Institutes, Inc., 4 Research Place, Suite 200, Rockville, Maryland 20850.

ISBN: 0-86587-483-2

Printed in the United States of America

SUMMARY CONTENTS

A. **Poultry Processing**
    Perdue Farms, Inc. 3

B. **Lower Extremities Evaluation**
    Dow Jones & Company 39

C. **VDTs/Workstations**
    US West Communications 51

D. **Manufacturing Evaluations**
    Harley-Davidson 111
    American Fuel Cell & Coated Fabrics Company 155
    Sancap Abrasives Company 225

E. **Vibration/Skin Temperature Evaluations**
    UNICCO 247

F. **Physiologic Measurements**
    Big Bear Grocery Warehouse 267

INDEX **319**

DETAILED CONTENTS

A. Poultry Processing

Perdue Farms, Inc. 3
Summary 3
Keywords 3
Introduction 4
Background 4
    Workforce 4
    Production Process 4
Methods 5
    Period Prevalence Rates 5
        Selection Criteria 5
        Case Definition 6
    Point Prevalence Rates 6
        Selection Criteria 6
        Case Definition 6
        Tension Neck Syndrome 6
        Rotator Cuff Tendonitis 6
        Lateral Epicondylitis 7
        Medial Epicondylitis 7
        Carpal Tunnel Syndrome 7
        Tendonitis of the Wrist or Fingers 7
        Non-Specific Proximal Interphalangeal (PIP) Joint Dysfunction 7
    Medical Management and CTD Prevention Program 7
Results 7
    Lewiston 7
        Prevalence Rates 7
        Medical Management 9
        CTD Prevention Program 10
    Robersonville 11
        Prevalence Rates 11
        Medical Management 12
        CTD Prevention Program 13
Discussion 14
    Lewiston 14
        Prevalence Rates 14
        Medical Management and CTD Prevention Program 16
    Robersonville 17
        Prevalence Rates 17
        Medical Management 19
    Overall Remarks 20
Recommendations 20
    Engineering Controls 20
    Administrative Controls 21
    Medical Controls 21
    Other 21
References 25
Authorship and Acknowledgements 25
Distribution and Availability 26
HETA 89-307-Table 1, Lewiston - Participation - All Employees 28

Table of Contents

HETA 89-307-Table 2, Lewiston - Participation - Women ... 28
HETA 89-307-Table 3, Age, Race, Gender and Length of Employment by Exposure Group - Lewiston - All Participants ... 29
HETA 89-307-Table 4, Age, Race and Length of Employment by Exposure Group - Lewiston - Women ... 29
HETA 89-307-Table 5, CTD Symptoms, Period Prevalence Cases and Point Prevalence Cases - Lewiston - All Participants ... 30
HETA 89-307-Table 6, CTD Symptoms, Period Prevalence Cases and Point Prevalence Cases - Lewiston - Women ... 31
HETA 89-307-Table 7, Participation - Robersonville- All Employees ... 32
HETA 89-307-Table 8, Participation - Robersonville-Women ... 32
HETA 89-307-Table 9, Age, Gender and Length of Employment by Exposure Group - Robersonville - All Participants ... 33
HETA 89-308-Table 10, Age, Race and Length of Employment by Exposure Group - Robersonville - Women ... 33
HETA 89-307-Table 10, CTD Symptoms, Period Prevalence Cases and Point Prevalence Cases, Robersonville - All Participants ... 34
HETA 89-307-Table 12, CTD Symptoms, Period Prevalence Cases and Point Prevalence Cases - Robersonville - Women ... 35
Appendix A - Surveillance Questionnaire ... 36
Appendix B - Upper Extremity (UE) CumulativeTrauma Disorders (CTD) Algorithm ... 38

B. Lower Extremities Evaluation

Dow Jones & Company ... 39
Summary ... 39
Keywords ... 40
Introduction ... 41
Background ... 41
Job Descriptions ... 42
Methods ... 42
    Medical ... 42
    Ergonomic ... 43
Evaluation Criteria ... 43
Results and Discussion ... 44
    Medical ... 44
    Demographic and Work History Characteristics ... 44
    Lower Extremity Discomfort ... 44
    Ergonomic ... 45
Conclusions ... 46
Recommendations ... 46
References ... 48
Authorship and Acknowledgements ... 48
Distribution and Availability of Report ... 49
Table 1, Dow Jones & Company, Demographic and Work History Characteristics ... 50

C. VDTs/Workstations

US West Communications ... 51
Summary ... 51

Table of Contents

Keywords 54
Introduction 55
Background 55
Methods 56
    City and Participant Selection 56
    Musculoskeletal Outcomes 56
    Demographics and Individual Factors 57
    Work Practices and Work Organization 57
    Psychosocial 58
    Electronic Performance Monitoring 58
    Keystoke Information 58
    Physical Workstation and Postural Measurements 59
    Statistical Analysis 59
Results 60
    Musculoskeletal Disorders 61
    Demographic and Individual Factors 62
    Work Practices and Organization Characteristics 62
    Psychosocial 63
    Electronic Performance Monitoring 63
    Keystrokes 64
Discussion 64
    Potential Work-Related Musculoskeletal Disorders 64
    Other Limitations 67
    Demographics 68
    Individual Factors 68
    Work Practices and Work Organization 69
    Psychosocial 70
    Electronic Performance Monitoring 72
    Keystrokes 72
    Physical Workstation 73
    Statistical 73
Conclusions 73
Recommendations 73
References 74
Authorship and Acknowledgements 75
Distribution and Availability of Report 82
HETA 89-299-Table 1, Physical Examination Criteria for Various Medical Conditions 83
HETA 89-299-Table 2, Criteria for Potential Work-Related Cases of Musculoskeletal Disorders 85
HETA 89-299-Table 3, Cumulative Symptoms Score Calculation for the Neck 86
HETA 89-299-Table 4, Participation Rates 86
HETA 89-299-Table 5, Prevalence of Potential Work-Related Upper Extremity Musculoskeletal
    Disorders (UE Disorders) by City and Job Title 87
HETA 89-299-Table 6, Types of Musculoskeletal Conditions Identified on the Physical
    Examination 87
HETA 89-299-Table 7, Prevalence of Potential Work Related Neck Musculoskeletal Disorders
    (Neck Disorders) by City and Job Title 88
HETA 89-299-Table 8, Prevalence of Potential Work-Related Shoulder Musculoskeletal
    Disorders (Shoulder Disorders) by City and Job Title 89
HETA 89-299-Table 9, Prevalence of Potential Work-Related Elbow Musculoskeletal Disorders
    (Elbow Disorders) by City and Job Title 89
    90

Table of Contents

HETA 89-299-Table 10, Prevalence of Potential Work-Related Hand-Wrist Musculoskeletal Disorders (Hand-Wrist Disorders) by City and Job Title 90
HETA 89-299-Table 11, Associations With Potential Work-Related Upper Extremity Musculoskeletal Disorders in the Logistic Regression Model 91
HETA 89-299-Table 12, Associations With Potential Work-Related Upper Extremity Musculoskeletal Symptoms in the Linear Regression Model 92
HETA 89-299-Table 13, Prevalence of Physician- Diagnosed Medical Conditions 93
HETA 89-299-Table 14, Work Practice Characteristics 93
HEAT 89-299-Table 15, Work Organization Characteristics 93
HEAT 89-299-Table 16, Psychosocial Variables 95
HEAT 89-299-Table 17, Electronic Performance Monitoring Variables 96
HEAT 89-299-Table 18, Monitoring VariableAssociations With Potential Work-Related Upper Extremity Musculoskeletal Symptoms in the Linear Regression Model 97
Figure 1, Outline of Statistical Analysis 98
Appendix A, Psychosocial Variables 99
Appendix B, Electronic Performance Monitoring Variables 100
Appendix C, Physical Workstation Characteristics and Postural Measurements 101
Table C-1, Physical Workstation Characteristics 103
Table C-2, Physical Workstation Characteristics - Postural Variables 105
Table C-3, Physical Workstations Characteristics - Chair 106
Table C-4, Physical Workstation Characteristics - Keyboard 106
Table C-5, Physical Workstations Characteristics - Screen 107
Table C-6, Physical Workstation Characteristics - Postural Variables 107
Figure C-1 (Appendix C) Posture Measurements 108
Figure C-2 (Appendix C) Posture Measurements 109

D. Manufacturing Evaluations

Harley-Davidson 111
Summary 111
Keywords 112
Background 113
    Work-Related Musculoskeletal Disorders 113
    Workforce 114
    Process Description 115
Design and Methods 116
    Ergonomic Evaluation 116
        Initial Evaluation 116
        Follow-Up Evaluations 117
Results 118
    Ergonomic Evaluation 118
        Milling 118
        Hand-Arm Vibration Exposure 119
        Truing (Assembly and Centering) 119
        Balancing 120
    Rates of Musculoskeletal Disorders 121
Discussion 122
Recommendations 124
    Engineering Controls 124
        Flywheel Milling 124
        Assembly and Truing Flywheels 125

Table of Contents

        Flywheel Balancing 125
      Work Practices 125
        Flywheel Milling 125
        Flywheel Balancing 125
      Organizational 125
      Medical Surveillance 126
      Other 126
References 127
Authorship and Acknowledgements 130
Distribution and Availability 130
Table 1, Evaluation of Ergonomic Changes 131
Table 2, Comparison of Pre-and Post-Interventions of Manual Handling of Flywheel Milling 133
Table 3, Steps in the Flywheel Deburring MachinePurchase 134
Table 4, Calculations Using 1991 NIOSH Lifting Formula For Flywheel Truing
    Machine Operator Task 1 136
Table 5, Calculations Using 1991 NIOSH Lifting Formula For Flywheel Truing
    Machine Operator Task 2 137
Table 6, Calculations using 1991 NIOSH Lifting Formula for Flywheel Assembly Lift for
    Balancing Job 138
Table 7, Entire Production Facility 139
Table 8, Flywheel Department 139
Figure 1, Flywheel Milling Job #1 140
Figure 2, Flywheel Milling Job #2 140
Figure 3, Worker in Flywheel Milling Job Using A Hand-Held Grinder 141
Figure 4, Worker in Flywheel Milling Positioning A Flywheel on Deburring
    Machine Robot Arm 142
Figure 5, Flywheel Truing Operations Using 5-Pound Brass-Head Hammer 143
Figure 6, Flywheel Truing Operations Using 40 Ton Press 144
Figure 7, Worker Lifting Flywheel Assembly From Tote Bin To Truing Press 145
Figure 8, Worker Lifting Flywheel Assembly From Truing Press 146
Figure 9, Worker Placing Flywheel Assembly On Worktable (Truing Area) 147
Figure 10, Worker Placing Flywheel Assembly In Tote Bin (Truing Area) 148
Figure 11, Worker Placing Flywheel In Balancing Machine 149
Figure 12, Overhead Hoist Moving Flywheel To Balancing Machine 150
Figure 13, Employment Growth an Rates of Musculoskeletal Disorders 151
Appendix A, NIOSH Lifting Equation 152
Appendix A, Table 1, Frequency Multiplier (FM) 153

American Fuel Cell and Coated Fabrics Company **155**
Summary 155
Keywords 155
Introduction 157
Background 158
Evaluation Criteria 158
    General 160
Methyl Chloroform, Methyl Ethyl Ketone, Ethanol 160
    Ergonomics 161
    Heat Stress 161
    Confined Spaces 162
Evaluation Design 164

Table of Contents

    Air Monitoring 164
    Ventilation Assessment 164
    Heat Stress 164
    Medical 165
    Ergonomics 165
Results 166
    Methyl Ethyl Ketone 166
    1,1,1-Trichloroethane 166
    Ethanol 166
    Heat Stress 167
    Medical 167
    Ergonomics 168
    Fittings Department 168
    Rubber Cutting Department 172
    Innerliner and Outerply Departments 172
    Nylon Spray Department 173
    Cleaning/Inspection and Repair Departments 174
    Onion Tank Assembly 174
Discussion and Conclusion 175
    Solvent Exposures 175
    Respiratory Protection 175
    Confined Spaces 176
    Ergonomics 177
    Medical 178
    Heat Stress 178
Recommendations 178
    Ergonomics 183
    Work Practices 184
    Confined Spaces 184
    Respiratory Protection 185
    Heat Stress 185
    Medical 186
    Ventilation 188
References 191
Authorship and Acknowledgements 192
Distribution and Availability of Report 192
HETA 90-246-Table 2, Toxicity and Permissible Exposure Information, Amfuel, Inc. 194
HETA 90-246-Table 4, Estimated Metabolic Rate - Final Finish Operator, Amfuel,
    Magnolia, Arkansas 195
HETA 90-246-Table 5, Confined SpaceClassification Table, Amfuel, Magnolia, Arkansas 196
HETA 90-246-Table 6, Check List of Considerations for Entry, Working In and
    Exiting Confined Spaces, Amfuel, Magnolia, Arkansas 197
HETA 90-246-Table 7, Sampling and Analytical Methods, Amfuel, Magnolia, Arkansas 198
HETA 90-246-Table 8, Results From Personal Breathing-Zone and General Area Air Samples
    for Methyl Ethyl Ketone 199
HETA 90-246-Table 8, Results From Personal Breathing-Zone and General Area Air Samples
    for Methyl Ethyl Ketone, Amfuel, Magnolia, Arkansas 200
HETA 90-246-Table 8, Results from Personal Breathing-Zone and General Area Air Samples
    for Methyl Ethyl Ketone, Amfuel, Magnolia, Arkansas 201

Table of Contents

    HETA 90-246-Table 8, Results From Personal Breathing-Zone and General Area Air Samples
        for Methyl Ethyl Ketone, Amfuel, Magnolia, Arkansas 202
    HETA 90-246-Table 8, Results From Personal Breathing-Zone and General Area Air Samples
        for Methyl Ethyl Ketone 203
    HETA 90-246-Table 9, Results From Personal Breathing Zone and General Area Air Samples
        for 1,1,1-trichloroethane, Amfuel, Magnolia, Arkansas 204
    HETA 90-246-Table 9, Results From Personal Breathing and General Area Air Samples for
        1,1,1-trichloroethane, Amfuel, Magnolia, Arkansas 205
    HETA 90-246-Table 9, Results From Personal Breathing Zone and General Area Air Samples
        for 1,1,1-trichloroethane, Amfuel, Magnolia, Arkansas 206
    HETA 90-246-Table 10, Results From Air Samples for Ethanol, Amfuel,
        Magnolia, Arkansas 207
    HETA 90-246-Table 11, Results of Heat Stress Monitoring (°F), American Fuel
        and Cell and Coated Fabrics Company, Magnolia, Arkansas, August 20, 1991 208
    HETA 90-246-Table 12, Results From Personal Heat Stress Dosimetry, Amfuel,
        Magnolia, Arkansas 209
    HETA 90-246-Appendix A, Elements of a Comprehensive Heat Stress Management Program .. 210
    Appendix B, Selected Local Exhaust Ventilation Designs 213
    Figure 1 - NIOSH Recommended Exposure Limits 214
    HETA 90-246, Figure 2 - Medical Interview Data, Amfuel Corporation 215

Sancap Abrasives, Inc. 225
    Summary 225
    Keywords 225
    Introduction 226
    Background 226
    Work Description 227
    Evaluation Criteria 228
        Cumulative Trauma Disorders 228
        Low Back Pain 230
    Evaluation Methods 230
    Results and Discussion 231
        Medical Evaluation 231
        Ergonomic Evaluation 232
        Safety Issues 234
    Recommendations 235
        Engineering Controls 235
        Administrative Controls 236
        Work Practices 236
    References 237
    Authorship and Acknowledgements 240
    Distribution and Availability of Report 241
    HETA 92-001, Table 1, Description Used by AT-6 Press Operators, Sancap Abrasives, Inc. 242
    HETA 92-001, Figure 1, Work Sequence at AT-6 Press, Sancap Abrasives, Inc. 243
    HETA 92-001, Figure 2, Biomechanical Evaluation, University of Michigan, 2D Static Strength
        Prediction Program, Sancap Abrasives, Inc. 245

E.   *Vibration/Skin Temperature Evaluations*

    UNICCO 247

Table of Contents

Summary 247
Keywords 247
Introduction 248
Background 248
    Facility Description 248
    Process Description 248
Evaluation Procedures 249
    Ergonomic 249
    Medical 251
Results and Discussion 251
    Ergonomic 251
    Medical 253
Conclusions 254
References 255
Authorship and Acknowledgements 255
Report Distribution and Availability 256
Table 1, Summary Results of Biomechanical Analysis 257
Table 2, Measurement of Average Direct Vibration (M/sec^2) 258
Table 3, Measurement of Average Indirect Vibration (M/sec^2) .. 258
Figure 1, Sketch of the BPVC Showing the Approximate Dimensions ... 259
Figure 2, Sketch of Horizontal Cut Through the L5/S1 Intervertebral Joint Showing Internal
    and External Biomechanical Forces 260
Figure 3, Free Body Diagram Describing Upright Posture Without BPVC ... 261
Figure 4, Free Body Diagram Describing Upright Posture With BPVC ... 261
Figure 5, Free Body Diagram Describing 6.5 Degree Forward Flexion Adjustment In
    Posture to Compensate for BPVC 261
Figure 6, Video Image of Worker Wearing the BPVC 262
Figure 7, Free Body Diagram Describing 85 Degree of Forward Flexion Without BPVC ... 263
Figure 8, Free Body Diagram Describing 85 Degree Flexion With BPVC ... 263
Figure 9, Schematic Diagram Showing Placement of the Accelerometers for the Direct
    Vibration Test 264
Figure 10, Frequency Weighting Curves Provided by ISO 264
Figure 11, Graphical Representation of Results of Skin Temperature Test ... 265

F.   *Physiologic Measurements*

Big Bear Grocery Warehouse 267
Summary 267
Keywords 269
Introduction 270
Background 270
    Plant and Job Description 270
        Workforce 270
        Job Activity 271
        Job Cycle 271
        Incentive System 271
    Incidence and Costs 272
Evaluation Design and Methods 272
    Medical Evaluation Methods 272
        OSHA 200 Logs and Workers' Compensation 272
        Questionnaire 272

Table of Contents

Ergonomic Assessment	274
Risk Factor Identification	274
Established Criteria	274
Heat Stress Evaluation	276
Results	279
Medical	279
OSHA 200 Logs and Workers' Compensation Data	279
Questionnaire	279
Ergonomic	281
Biomechanical Data	281
Metabolic Data	284
Workplace Layout Analyses	284
Heat Stress	285
Conclusions	287
Limitations	288
Recommendations	289
Ergonomic	289
Heat Stress	290
References	290
Authorship and Acknowledgements	294
Distribution and Availability	295
Table 1, OSHA 200 Log Entries for Selectors	296
Table 2, Demographics of Study Participants	296
Table 3, Injuries and Missed Workdays Reports by Questionnaire	297
Table 4, Rate of Reported Discomfort	297
Table 5, Summary Results of NIOSH Lifting Equation Evaluation	298
Table 6, Summary Results of 3-D Biomechanical Analysis	298
Table 7, Lifting Task Conditions by Origin and Description	299
Table 8, Dynamic Lifting Task Analysis	299
Table 9, Summary Table For Metabolic Criteria	300
Table 10, Heat and Humidity Measures on 5/14/92	301
Table 11, Heat and Humidity Measures on 7/14/92	302
Table 12, Metabolic Heat Production Rates by Task Analysis	303
Table 13, Estimated Metabolic Rate, Grocery Selectors	304
Figure 1, Examples of Listing Tasks Analyzed Using Revised NIOSH Equation	305
Figure 2, Body Map Discomfort Diagram	306
Figure 3, NIOSH Recommended Exposure Limits	306
Appendix A, Worker Questionnaire	307
Appendix B, NIOSH Lifting Equation Calculations	312
Appendix B, Table 1, Frequency Multiplier (FM)	313
Appendix B, Table 2, Coupling Multiplier	314
Appendix C, Acute Physical Effects Caused by Excessive Heat Stress	315
Appendix D, Elements of a Comprehensive Heat Stress Management Program	316
INDEX	**319**

INTRODUCTION

The topic of ergonomics became a focus of national attention after OSHA discovered a series of record keeping violations at some meatpacking plants, and levied unprecedented high fines against them. In the late 1980s, the meatpacking industry's incidence of disorders due to "repeated trauma" was approximately seventy-five times that of industry as a whole. (1) The violations produced settlement agreements with OSHA, which allowed these companies to enter into long-term ergonomic programs as a means of reducing their fines. The primary elements of these programs were worksite analysis, hazard prevention and control, medical management, and training. There was also a unique feature of these settlement agreements: they included a provision that a grant be made to NIOSH to study repetitive motion illnesses. NIOSH then developed a project called "participatory approach," which demonstrated the process of developing ergonomic teams comprised of front-line workers and supervisors to effect job changes. An analysis of this approach was subsequently published. (2)

The principal goal of an ergonomic program is to establish an ongoing mechanism for systematically identifying affected workers and jobs, implementing medical and work interventions, and evaluating the effectiveness of those interventions. (3) In order to be effective, an ergonomic program should have the support of management, engineering, safety, medical, and labor professionals. One mechanism used to achieve this goal is the establishment of an ergonomic team or committee to evaluate health data and determine priorities for job analysis and intervention, to evaluate job analysis data and select intervention strategies, and to evaluate program progress.

The Health Hazard Evaluation (HHE) program, designed to serve NIOSH's charge to respond to requests for the investigation of workplace hazards, has provided many important insights into workplace problems. A typical HHE involves studying the specific aspect of a job or a workplace, such as a manufacturing facility, firing range, lead paint removal project, or office building. Most HHEs occur as a result of employee concerns over symptoms or illnesses. NIOSH must then determine if these effects are workplace-related and, if so, what the causative agents are. NIOSH's HHEs not only help industries and labor sort through perplexing safety and health problems, but they also develop worker exposure data that is essential to both present and future research and development. NIOSH is conducting research and field testing methods to prevent workplace problems across the wide span of industries where they are occurring, such as in the meatpacking, poultry, agriculture, construction, retail, publishing, automotive repair, and data processing industries. Epidemiological methods are often used to identify jobs, tools, areas, plants, or industries with excessive risk. As an example, statistics such as incidence and prevalence rates and relative and attributable risk ratios are frequently used to identify high risk groups. NIOSH's medical evaluations typically involve reviews of the OSHA 200 logs, pertinent medical records, and confidential interviews with employees. Physiological testing may also be performed. Interviews often focus on work history, work-related cumulative trauma disorders (CTDs) and musculoskeletal symptoms, medical treatment, concerns related to the job, and suggestions for improving work conditions. While this publication emphasizes the use of studies where overexposures have been identified so that readers may benefit from the recommendations for controls, this should not be inferred to represent all workplaces for any given industrial operation.

Between 1982 and 1992, the reported number of musculoskeletal disorders of the upper extremities has steadily increased. In 1992, these disorders accounted for more than 60 percent of all occupational illnesses. Depending on the type of job being performed, these disorders may cause pain, restricted motion, and weakness in the hands, arms, shoulders, neck, back, and lower limbs. In addition, pain in the back, hips, and lower limbs caused by mechanical stresses on the lumbar spine presents one of the most common and costly problems in occupational health. It has been estimated that the lifetime prevalence of moderate to severe low back pain in the general population may be as high as 70 percent. (4) Work-related musculoskeletal disorders are responsible for escalating costs of workers' compensation claims as well as of diagnosis and treatment. The total amount of compensable costs to the nation for these disorders is estimated to exceed $20 billion annually. (5,6)

Disability resulting from carpal tunnel syndrome and other CTDs, also known as repetitive strain injuries, is the single fastest growing public health problem among American workers, according to NIOSH. Carpal tunnel is characterized by pain, numbness and tingling in the first three fingers, resulting from compression of the median nerve as it passes through the wrist. Such compression of the median nerve may occur when the finger flexor tendons, which also pass through the rigid carpal tunnel, become inflamed. (1) Although carpal tunnel syndrome is the most commonly diagnosed nerve entrapment disorder, it occurs much less frequently than other musculoskeletal disorders, such as tendinitis.

CTDs can be caused by any combination of personal and work factors. Personal factors include rheumatoid arthritis, endocrine disorders, acute trauma, vitamin B6 deficiency, wrist size and shape, obesity, gender, age, oral contraception use, and gynecological surgery. (3) Commonly identified physical-risk work factors include repeated or sustained exertions, forceful exertions, contact stresses, low temperature and vibration, lack of adequate rest or recovery, and certain stressful postures. (8) Sustained static loading (when the muscles are held in fixed positions for prolonged periods of time) and asymmetrical loading of the joints (which occurs during wrist extension and flexion, ulnar and radial wrist deviation, and shoulder abduction) are among the components of stressful postures. The result is impaired circulation to the involved tendons; tendon degeneration can follow. (9,10) Even a low-level load can increase the risk of musculoskeletal injury if the load is maintained for an extended period. (11) It is particularly important to ensure adequate recovery time from static exertions, as static work can be more fatiguing than dynamic work, even if the static work is lighter. (12,13)

In addition to physical-risk work factors, several psychosocial and work- organization characteristics of jobs have also been associated with musculoskeletal problems. These include working under the pressure of deadlines, lack of control over various job aspects, high workload without adequate recovery time, and a perceived lack of support from supervisors. (14) Therefore, in many respects it is difficult to separate the results of physical stress from psychological stress in jobs. But in order to develop adequate controls for a problem, both issues need to be addressed. For example, NIOSH has evaluated the impact of electronic performance monitoring on the operators of video display terminals (VDTs) along with other factors such as gender, use of bifocals, and work station design. (15) The result is that, since symptoms are often poorly localized, nonspecific, and episodic, and there is frequently more than one causal factor, these work-related musculoskeletal disorders, including low back pain, are often under-reported.

Repetitive (or stereotypical motions) are also known to result in fatigue and micro trauma to soft tissues and joints of the body. As the repetition rate increases and joint and tissue stress accumulates, the risk for musculoskeletal injury increases. Forceful movements are also linked to the development of musculoskeletal injuries. As muscle exertion increases, blood flow to the muscles decreases, resulting in fatigue. The harmful effects of force are exacerbated if combined with repetitive movements and stressful postures. One source of confusion is that fatigue and CTDs are often considered to be the same; however, while there are similarities between the two, there are also important distinctions. The primary distinction is that fatigue onset and recovery occurs within a much shorter period of time—minutes or hours rather than weeks, months, or years. On the other hand, frequent fatigue is a sign that long-term damage may be occurring.

Low back pain can be caused by direct trauma to the spine or a single "overexertion" of the muscles and ligaments, or it can result from multiple strains (i.e., repetitive trauma). (16) Lifting is a major cause of work-related back pain and impairment. Repeated lifting can cause pathological degeneration of the discs which lie between the vertebrae, causing damage to the spine, nerve irritation, and pain. Bending, stooping, pushing/pulling and prolonged sitting can also

Introduction

contribute to the onset of back injury. Also, there is some evidence that work-related psychological stress and lifestyle factors may increase the likelihood of back pain and the subsequent risk of prolonged disability. (17)

While most musculoskeletal studies focus on upper-body disorders, knee and other lower-limb injuries also can be caused by undue mechanical stresses on the knee joint, imposed by activities which require prolonged squatting or kneeling. Previous research has shown that postures which require near-maximum knee flexion can produce shear forces sufficient to stretch or otherwise damage knee ligaments. (18) Furthermore, squatting and kneeling have both been associated with a variety of nerve compression disorders. (19,20) NIOSH has evaluated this type of condition in several occupations including printing. (21)

Not all musculoskeletal disorders are associated with manual handling tasks in heavy industries. For example, caring for children is often not perceived as producing lifting/handling difficulties because children are generally considered to be "lightweight". However, increasing complaints of chronic fatigue and low back pain have been reported among nursery school teachers in Japan, and it has been noted that caring for children can lead to postures which impose heavy static loads on the musculoskeletal system. (22,23,24) Tasks such as lifting children from floor level may require muscle strengths in excess of the capabilities of this largely female worker population. NIOSH has evaluated this type of situation for teachers of young children. (25)

NIOSH's ergonomic studies provide an opportunity to view applications of tools such as the Two Dimensional (2D) Static Strength Predictor Program developed by the Center for Ergonomics at the University of Michigan. (26) This two dimensional static strength software program determines the biomechanical and static strength capabilities of the employee in relation to the physical demands of the work environment. The program can be used to evaluate the physical demands of a prescribed job and the merits of proposed workplace design and redesign prior to actual construction or reconstruction of the workplace or task.

Another application used in HHEs is the NIOSH Lifting Equation. (27) NIOSH first developed an equation in 1981 to evaluate manual lifting demands. Because this first equation could be applied to only a limited number of lifting tasks (sagital lifting tasks), it was revised and expanded in 1991 so that it could be applied to a larger percentage of lifting tasks. (28) The revised procedure can be used to evaluate asymmetrical lifting tasks, objects with less than optimal hand-container couplings, and a larger range of work durations and lifting frequencies. Three criteria were used to develop the equation: biomechanical, physiological and psychophysical. The biomechanical criterion takes into account the effects of lumbrosacral stress, which is most important in infrequent lifting tasks. The physiological criterion evaluates metabolic stress and fatigue associated with repetitive lifting tasks. Finally, the psychophysical criterion measures the workload based on the workers' perceptions of their lifting capabilities.

As noted earlier, NIOSH has developed guidelines for setting up ergonomic programs that incorporate worker participation. The primary issues to address are management commitment, training, team or committee composition, information sharing, activities and motivation and a process for evaluating performance. Table 1 further elaborates on key points in each of these issues. (2)

This approach has been examined by NIOSH and found to work as long as there is strong in-house direction and support and significant staff expertise in both team building and ergonomics. Furthermore, team size should be kept to a minimum, but should include production workers engaged in the jobs to be studied, area supervisors, and maintenance and engineering staff who can effect proposed job improvements. Goals for the program need to be realistic and take account of the possibility that solutions to some problems may not be immediately forthcoming.

Effective team problem solving requires sharing information which bears on the issues under study. Such information might include company statistics on CTD prevalence, worker compensation claims and costs, sickness, absenteeism, and employee turnover to assist in defining problem jobs. There should also be an opportunity to collect other data reflecting risk factors and to perform worker interviews.

A means for evaluating team and company efforts and results needs to be written into the overall plan for the ergonomics program. An example of such an approach is the program developed by Harley Davidson for the flywheel department of their motorcycle manufacturing operation in Milwaukee, WI. (29). The ergonomic committee was able to develop a means to lighten flywheel castings as a result of improved die-cast specifications, product flow, and better milling machines. The need to use brass hammers was eliminated, and an overhead lift was substituted for the manual handling of the assembled flywheel unit. As a result, the number and the severity rates of musculoskeletal disorders decreased.

Since engineers are generally not trained in ergonomics, it is not uncommon for teams to be addressing existing design problems that need to be retrofitted. Rather than assigning blame, companies need to take responsibility for training and careful review in the design stage, which can provide long term benefits.

As of this writing, it appears that NIOSH has been funded for the next two years, despite the current political climate in which, it has been suggested, next year's elections could have a significant impact on the agency's budget. The loss of NIOSH's contribution to the safety and health profession would leave a gap that academians and our professional organizations cannot fill. Most of us rely on the NIOSH Manual of Analytical Methods or the NIOSH Pocket Guide to Chemical Hazards as well as on many of the other documents that NIOSH produces. NIOSH supports university-based Educational Resource Centers which have developed highly trained scientists who have made the U.S. the world's leader in health and safety technology. In addition, NIOSH is not only a primary source of funding for many academic research studies in health and safety, but conducts research that is directly responsible for developing work practices and guidelines now considered essential for protecting worker health and safety. In the process NIOSH has developed an incredible wealth of case studies, examples of which are included in this book.

In its entire 57-year history, the American Conference of Governmental Industrial Hygienists (ACGIH) has never taken a position on a political issue. However, ACGIH recently broke with tradition and sent a letter to the members of the Senate Appropriations Committee to indicate the Conference's alarm about congressional discussions which could have devastating effects on the American work force's health and productivity. This is not the time for the safety and health community to rest. Remind your congressmen and senators often just how valuable NIOSH is. The U.S. worker cannot afford to loose this important resource.

Shirley A. Ness, CIH, CSP
November 6, 1995

Introduction

References

1. Sheridan, P.J.: Meatpackers move to cut injury rates. *Occupational Health*, May 1991, pp. 81-85.

2. Gjessing, C.C., Schoenborn, T.F., Cohen, A. (eds): *Participatory ergonomic interventions in meatpacking plants.* U.S. Dept. of Health and Human Services, NIOSH, Cincinnati, OH, 1994. DHHS (NIOSH) Pub. No. 94-124.

3. Armstrong, T.J.: *Ergonomics guides: An ergonomics guide to carpal tunnel syndrome.* AIHA, 1983.

4. Frymoyer, J.W., Pope, M.H., Clements, J.H., Wilder, D.G., McPherson, B. and Ashikaga, T.: Risk factors in low-back pain. *The Journal of Bone and Joint Surgery* 65A: 213-218, 1983.

5. Webster, B.S. and Snook, S.H.: The cost of 1989 workers' compensation low back pain claims. *Spine* 19(10): 1111-1116, 1994a.

6. Webster, B.S. and Snook, S.H.: The cost of compensable upper extremity cumulative trauma disorders. *JOM* 36(7): 714-717, 1994b.

7. Phalen, G.S.: The carpal tunnel syndrome, clinical evaluation of 598 hands. *Clin. Orthop. Rel. Res.* 83: 29-40, 1972.

8. Putz-Anderson, V.: *Cumulative trauma disorders: A manual for musculoskeletal disorders of the upper limbs.* Philadelphia: Taylor & Francis. 1988.

9. Hagberg, M.: Occupational musculoskeletal stress and disorders of the neck and shoulder: A review of possible pathophysiology. *International Arch. Env. Hlth* 53: 269-278, 1984.

10. Herberts, P. and Kadefors, R.: A study of painful shoulder in welders. *Acta Orthop Scand* 47: 381-387, 1976.

11. Westgaard, R., Waersted, M., Jansen, T., and Aaras, A.: Muscle load and illness associated with constrained body postures. In: Corlett, N., Wilson, J., Manenica, I., eds., *The Ergonomics of Working Postures*, vol. 1, Philadelohia: Taylor & Francis, pp. 5-18, 1986.

12. Malmqvist, R., Ekholm, I., Lindstrom, L., Petersen, I., Ortengren, R., Bjuro, T., et al.: Measurement of localized muscle fatigue in building work. *Ergonomics* 24: 695-709, 1981.

13. Corlett, E.N.: Analysis and evaluation of working posture. In: Kvalseth, T.O., ed., *Ergonomics of Workstation Design*, London: Butterworths, p. 13, 1983.

14. Wallace, M., and Buckle, P.: Ergonomic aspects of neck and upper limb disorders. In: Osborne, D.J., ed., *International Review in Ergonomics: Current Trends in Human Factors Research and Practice*, vol. 1, London: Taylor & Francis, 1987.

15. NIOSH Health Hazard Evaluation Report US West Communications, HHE 89-299-2230. 1991. In this volume.

16. Pope, M.H., Andersson, G.B.J., Frymoyer, J.W. and Chaffin, D.B.: Occupational low back pain: Assessment, treatment and prevention. *Mosby Year Book*, St. Louis, MO, 1992.

17. Biogos, S., Spengler, D.M., Martin, N.A., Zeh, J., Fisher, L. and Nachemson, A.: Back injuries in industry: A retrospective study. III. Employee-related factors. Spine 11: 252-256.

18. Ariel, B.G.: Biomechanical analysis of the knee joint during deep knee bends with heavy load. In: R.C. Nelson and C.A. Moorehouse, eds., *Biomechanics*. University Park Press, Baltimore, MD.

19. Spans, F.: Occupational nerve lesions. In: P.J. Vinken and G.W. Bruyn, eds., *Handbook of Clinical Neurology*, vol.7: Diseases of nerves, Part 1. Elsevier, NY.

20. Aguayo, A.J.: Neuropathy due to compression and entrapment. In: *Peripheral Neuropathy*, P.J. Dyck, P.K. Thomas, and P.K. Lambert, eds, W.B. Saunders, Philadelphia, 1975.

21. NIOSH Health Hazard Evaluation Report: Dow Jones and Co., Inc., HHE 90-251-2128. 1993. In this volume.

22. Nishiyama, K., Sato, K., Yondo, Y., Nakaseko, M., Hosokawa, M. and Tokunaga, R.: Work and work load of nursery teachers in institutions for mentally and physically handicapped children. *Arhiv. Az Higijenu Rada I Toksikologiju* 30 (suppl): 1235-1242, 1979.

23. Kumagai, S., Nakachi, S., Hanaoka, M., Kataoka, A. and Shibata, T.: Work load of nursery teachers in a nursery school: Relationship between age of children and work load. *Jap. J. Ind. Hlth* 32(6): 470-477, 1990.

24. Corlett, E.N., Lloyd, P.V., Tarling, C., Troup, J.D.G. and Wright, B.: *The guide to the handling of patients, 3rd ed*. National Back Pain Assoc, Middlesex, UK, 1992.

25. NIOSH Health Hazard Evaluation Report: Terry's Montessori School HHE 93-995-2442. 1994.

26. NIOSH Health Hazard Evaluation Report: Sancap Abrasives Inc. HHE 92-1-2444. 1994. In this volume.

27. NIOSH Health Hazard Evaluation Report: Yorktown Inc. HHE 88-384-2062. 1991.

28. Waters, T.R., Putz-Anderson, V., Garg, A. and Fine., L.J.: Revised NIOSH equation for the design and evaluation of manual lifting tasks. *Ergonomics* 36(7): 749-776, 1993.

29. NIOSH Health Hazard Evaluation Report: Harley Davidson Inc., Milwaukee, WI. HHE 91-208-2422. 1994. In this volume.

Introduction

TABLE I

Issue	Factors to Consider
Management Commitment	1. Top management's commitment and support of worker participation approaches to company problem-solving needs is critical as is the cooperation of lower level supervisors and union officials or recognized worker leaders.
	2. Policy declarations on the importance of participative approaches in addressing workplace issues require follow-up management actions to prove credibility. Those having merit are worker memberships on existing or newly-formed groups at various levels within the organization, including those that have authority to make decisions in local areas of operation, providing timely responses to worker-generated proposals for problem-solving and resources to implement solutions.
	3. Efforts will be needed to redefine the roles of mid-level supervisors as mentors to workers, to work with them in promoting ideas for work improvement and ways that they can be implemented.
Training	1. Workers and management staff plus others who may be formed into a work team, task group or committee will require additional training to ensure effective joint actions. workers will need training in communication skills and abilities to interact in group problem-solving tasks; managers in listening and feedback skills.
	2. Both workers and managers plus other participant members of a work team or task group should be given the necessary technical training to appreciate the targeted problems at issue. Resources for this and other add-on training should include provisions for outside consultants or experts as may be necessary.
	3. Training practices should stress active forms of instruction focused on issues relevant to the trainees experience. Special needs of those having language difficulties or other impediments to comprehension should be addressed.
Composition	1. No single form of worker participation can effectively fit all needs. Approaches depend upon the problem(s) to be addressed, whether limited to one group, area or operation or having broader ramifications, the abilities of the workforce involved, and the climate of the organization in terms of using participative approaches in problem-solving.
	2. Teams formed to address workplace problems which cut across different units in an organization should include representatives from all such groups in addition to impacted workers, management persons and technical consultants as needed. Groups of 7 to 15 persons can afford ample interactions and cohesiveness in actions.
	3. Precautions should be taken to prevent supervisors/managers, specialists, and consultants on a team from intimidating frontline worker members of a team or dominating discussion.
Information Sharing	1. Effective worker participation and team efforts to solve problems demand access to information germane to the issues in question.
	2. As the team participants may represent different operations and be at different staff levels, the success of group efforts can hinge on sharing information.
	3. Management must be up-front and honest in communicating their support for participative decision making and in acknowledging possible consequences of proposed actions. Worker concerns for job security are certain to raise questions.

NIOSH Case Studies in Ergonomics

Activities & Motivation	1. Team-building activities invariably include meetings to clarify aspects of the problem, doing data gathering and analyses to isolate causal or contributing factors, developing remedial suggestions and planned efforts at implementation. Procedures reflecting orderly, systematic ways for dealing with each of these elements offer the best chances for success.
	2. Goal-setting and frequent feedback to mark progress toward the goals in a group's problem-solving efforts are key ways for motivating performance.
	3. Team leader commitments to the objectives of the group can facilitate accomplishments.
	4. Management's recognition and rewards for team success in problem solving work can reinforce and sustain the continued interest of team members.
Evaluation	1. Team performance efforts need to be evaluated. suitable process and/or outcome measures should be used for that purpose.
	2. Surrogate indicators may offer alternatives to more basic measures in cases where the latter data do not satisfy conditions for meaningful evaluations.

Source: Gjessing, C.C., Schoenborn, T.F., Cohen, A. (eds): Participatory ergonomic interventions in meatpacking plants. U.S. Dept. of Health and Human Services, NIOSH, Cincinnati, OH, 1994. DHHS (NIOSH) Pub. No. 94-124.

NIOSH CASE STUDIES IN ERGONOMICS

HETA 89-307-2009
FEBRUARY 1990
PERDUE FARMS, INC.
LEWISTON, NORTH CAROLINA
ROBERSONVILLE, NORTH CAROLINA

NIOSH INVESTIGATORS
Stuart Kiken, M.D., M.P.H.
William Stringer, M.S.
Lawrence Fine, M.D., Dr.P.H.
Thomas Sinks, Ph.D.
Shiro Tanaka, M.D., M.P.H.

I. SUMMARY

In July, 1989, the National Institute for Occupational Safety and Health (NIOSH) received a request from the North Carolina Department of Labor, Division of Occupational Safety and Health, for technical assistance in evaluating cumulative trauma disorders (CTDs) of the neck and upper extremity among employees at two Perdue Farms, Inc. poultry processing plants, located in Lewiston and Robersonville, North Carolina. In response to this request, investigations were conducted at both facilities. The main objective of the investigations was to determine the prevalence of CTDs among employees working in selected departments, which, on the basis of a walk-through evaluation, had been characterized as having jobs with either higher exposure (HE) or lower exposure (LE) to repetitive and forceful motions and/or extreme and awkward postures of the upper extremity.

At Lewiston, 36% of the 174 employees who participated in NIOSH's survey had work-related CTDs in the last year as determined by questionnaire alone, and 20% had current work-related CTDs as determined by questionnaire and physical exam. Most CTDs involved the hand and/or wrist. Employees in HE departments were 4.4 times more likely than employees in LE departments to have CTDs by questionnaire (95% CI 1.48-13.24), and 3.6 times more likely by questionnaire and physical exam (95% CI 0.91-14.28). For hand and wrist injuries, the respective relative rates were 4.7 (95% CI 1.21-18.54) and 3.0 (95% CI 0.76-12.14).

At Robersonville, 20% of the 120 participants had significant work-related CTDs in the last year as determined by questionnaire alone, and 8% had current work-related CTDs as determined by questionnaire and physical exam. Most CTD's here also involved the hand and/or wrist. Employees in HE departments were 10.1 times more likely than employees in LE departments to have CTDs by questionnaire (95% CI 1.42-72.24).

Given the high employee turnover rates at the Lewiston and Robersonville facilities, the relative rates noted above may actually represent an underestimation of the true risks. (Workers with CTDs selectively tend to leave employment, resulting in a "survivor bias.")

On the basis of this investigation, NIOSH investigators concluded that a neck and upper extremity CTD hazard exists at the Lewiston and Robersonville facilities of Perdue Farms, Inc. Recommendations to prevent and manage these CTDs are provided in Section VII.

KEYWORDS: SIC 2016 (poultry processing plant), poultry, chicken, meatpacking, cumulative trauma disorder, carpal tunnel syndrome, tendonitis, epicondylitis, tension neck

Health Hazard Evaluation Report No. 89-307

II. INTRODUCTION

In July, 1989, the National Institute for Occupational Safety and Health (NIOSH) received a request from the North Carolina Department of Labor, Division of Occupational Safety and Health, for technical assistance in evaluating cumulative trauma disorders (CTDs) of the neck and upper extremity among employees at two Perdue Farms, Inc. poultry processing plants, located in Lewiston and Robersonville, North Carolina. Investigations were conducted (1) on July 10, 1989 at the Lewiston facility, and (2) from July 31, 1989 to August 3, 1989 at both facilities. The main objective of the investigations was to determine the prevalence of CTDs among employees working in selected departments, which, on the basis of a walk-through evaluation, had been characterized as having jobs with higher exposure (HE) or lower exposure (LE) to repetitive and forceful motions and/or extreme and awkward postures of the upper extremity. In addition, investigators assessed the medical management of injured workers and examined the company's CTD prevention program. Preliminary results were reported to the North Carolina Department of Labor, Division of Occupational Safety and Health, on September 11, 1989 and September 18, 1989.

III. BACKGROUND

A. Workforce

Perdue Farms, Inc. produces and packages boneless chicken products, chicken parts, and whole chickens for wholesale distribution. It is the fifth largest poultry processor in the United States, with four plants in North Carolina. The Lewiston facility employs approximately 2600 workers and processes over 420,000 chickens/day. The Robersonville facility employs approximately 550 workers and processes over 120,000 chickens/day. At each plant, there are two production shifts and one sanitation shift per day. The annual employee turnover rate is close to 50% at Lewiston and 70% at Robersonville.

B. Production Process

At the Lewiston facility, chickens are removed from the crates in which they are received and suspended by their feet on an overhead chain conveyor. They are then stunned (and tranquilized) by an electric shock, and their necks cut by machine (to kill and exsanguinate them). Machines and workers stationed along the conveyor (which moves at a pre-determined speed) subsequently defeather the chickens, remove their feet, and eviscerate the birds. The chickens are then graded (to determine which should be sold whole) and cooled in a water bath. Birds that are not sold whole or for parts are deboned and cut. Deboning is done by hand on either a conventional table deboning line or a cone deboning line. (A small percentage are deboned by machine on a newly installed, automated line.) Cutting is done by machine and by hand. The meat is then packed in styrofoam trays, weighed and labeled, and prepared for shipping.

Health Hazard Evaluation Report No. 89-307

At the Robersonville facility, the process is similar. The main differences are that at Robersonville (1) deboning is only done by hand, and (2) meat is packed in boxes containing ice instead of in styrofoam trays.

IV. METHODS

A. Period Prevalence Rates

Period prevalence rates for the 12 months prior to NIOSH's investigation were calculated from standardized questionnaires administered to current employees at the Lewiston and Robersonville facilities. The questionnaires elicited demographic information, medical and work histories, and data on neck and upper extremity CTDs.

1. Selection Criteria

A walk-through survey was conducted at the Lewiston facility on July 10, 1989 to select departments in which employees had jobs with either higher exposure (HE) or lower exposure (LE) to repetitive and forceful motions and/or extreme and awkward postures of the upper extremity. On the basis of this walk-through, the following departments were determined to contain HE jobs: receiving, evisceration, whole bird grading, cut up, and deboning. LE jobs were determined to be in the maintenance, sanitation, quality assurance, and clerical departments. (These categorizations were substantially corroborated by the North Carolina Department of Labor's analysis of the Lewiston facility's OSHA 200 Log for 1988 and 1989.)

For purposes of determining period prevalence rates, employees were then selected from these HE and LE departments. Analagous departments at the Robersonville facility were similarly categorized. (The LE jobs were selected because their CTD risk levels were expected to be lower than those of HE jobs. If the LE jobs were compared to totally nonrepetitive and nonforceful jobs, however, it is possible that some of them would also have elevated CTD risk levels.)

An attempt was made to select 70-80% of participants from HE departments and 20-30% from LE departments. At Lewiston, participants were randomly chosen from a roster of current employees that had been provided by Perdue Farms, Inc. At Robersonville (which has fewer employees), participants were similarly chosen from HE and LE jobs. For jobs in which there were small numbers of employees, all workers present were asked to participate.

Health Hazard Evaluation Report No. 89-307

2. Case Definition

An upper extremity CTD was said to exist if:

a. in the past year, the employee had pain, aching, stiffness, burning, numbness, or tingling in the neck, shoulder, elbow, wrist or hand; and

b. the symptoms began after employment at the plant; and

c. the symptoms were not due to an accident or injury that occurred outside of work; and

d. the symptoms lasted more than 8 hours, and occurred 4 or more times in the last year.

B. Point Prevalence Rates

Physical examinations were performed by NIOSH physicians (certified in internal medicine and/or occupational medicine) on employees who completed questionnaires. The examinations were of the neck and upper extremity, and included inspection, palpation, and the performance of various diagnostic maneuvers.

1. Selection Criteria

See IV. A. 1.

2. Case Definition

For purposes of determining point prevalence, an upper extremity CTD case was said to exist if an employee satisfied (1) the questionnaire case definition (described in IV. A 2.), and (2) the physical examination diagnostic criteria for a neck, shoulder, elbow, wrist or hand injury affecting the same area as the one identified on the questionnaire.

The diagnostic criteria are described below. For those tests requiring an assessment of pain, employees were asked to quantify their pain on a scale of 0 to 8 (with 0 representing no pain and 8 representing the worst pain experienced in one's life).

Tension Neck Syndrome

Tension neck syndrome was defined as pain ≥ 3 on two of the following tests: passive flexion, extension, lateral bending or rotation; resisted flexion, extension, lateral bending or rotation.

Rotator Cuff Tendonitis
Rotator cuff tendonitis was defined as pain ≥ 3 on active and resisted shoulder abduction.

Health Hazard Evaluation Report No. 89-307

Lateral Epicondylitis

Lateral epicondylitis was defined as pain ≥ 3 at the lateral epicondyle on resisted wrist extension.

Medial Epicondylitis

Medical epicondylitis was defined as pain ≥ 3 at the medial epicondyle on resisted wrist flexion.

Carpal Tunnel Syndrome

Carpal tunnel syndrome was defined as a positive Tinel's sign (pain, numbness, or tingling in the median nerve distribution resulting from light tapping over the proximal wrist crease) and a positive Phalen's sign (pain, numbness, or tingling in the median nerve distribution resulting from complete flexion of the wrist for 60 seconds).

Tendonitis of the Wrist or Fingers

Tendonitis of the wrist or fingers was defined as pain ≥ 3 on resisted flexion or extension of the wrist or fingers.

Non-Specific Proximal Interphalangeal (PIP) Joint Dysfunction

Non-specific PIP joint dysfunction was defined as decreased range of motion at the PIP joint.

C. Medical Management and CTD Prevention Program

The medical management of injured workers and the company's CTD prevention program were assessed (1) in the questionnaire and, (2) by reviewing Perdue Farms, Inc. "Repetitive Motion Disorders Action Plan" for Robersonville (dated 11/15/88).

V. RESULTS

A. Lewiston

1. Prevalence Rates

   a. Participation

   ### All Participants

   One hundred seventy-four (174) employees were interviewed and examined. (Only 3 employees who were asked to participate in the investigation refused.) Of these, 81.7% worked in higher exposure (HE) departments (receiving-9.8%, evisceration-9.8%, cut up-23.0%, and deboning-37.4%), and 18.3% worked in lower exposure (LE) departments

Health Hazard Evaluation Report No. 89-307

(maintenance-4.6%, quality assurance-6.3%, and clerical-7.5% [0.1% rounding error]). Two workers from the (whole bird) packing department and one from the stretch bag department were included in the HE group. (Table 1)

Women Participants

One hundred forty-two (142) employees were women. Of these, 86.0% worked in HE departments (evisceration-12.0%, cut up-28.2%, and deboning-43.7%), and 14.1% worked in LE departments (quality assurance-5.6% and clerical-8.5%). (0.1% rounding error) Two workers from the (whole bird) packing department and one from the stretch bag department were included in the HE group. (Table 2)

b. Demographics

All Participants

Eighty-one and six-tenths percent (81.6%) of participants were female and 18.4% were male. Eighty-nine and one-tenth percent (89.1%) were black and 10.9% were white. The mean age was 31 years, the mean length of employment 5.5 years, and the mean "time at current job" 47.0 months. The HE group had a greater percentage of women, blacks, and younger people than the LE group. (Table 3)

Women Participants

Ninety and eight-tenths percent (90.8%) of women were black and 9.2% were white. The mean age was 31 years, the mean length of employment 5.4 years, and the mean "time at current job" 44.4 months. The HE group had a greater percentage of blacks and younger people than the LE group. (Table 4)

c. Period Prevalence

All Participants

One hundred thirty-five (77.6%) employees reported symptoms compatible with neck or upper extremity CTDs over the one-year study period and 62 (35.6%) met the period prevalence case definition. Employees in the HE group were 2.3 times more likely to have symptoms (95% CI 1.47-3.63) and 4.4 times more likely to satisfy the period prevalence case definition (95% CI 1.48-13.24) than employees in the LE group. (Table 5)

Health Hazard Evaluation Report No. 89-307

Women Participants

One hundred nineteen (83.8%) women employees reported symptoms compatible with neck or upper extremity CTDs over the one-year study period and 54 (38.0%) met the period prevalence case definition. Employees in the HE group were 2.6 times more likely to have symptoms (95% CI 1.44-4.78) and 8.7 times more likely to satisfy the period prevalence case definition (95% CI 1.27-59.33)) than employees in the LE group. (Table 6)

d. Point Prevalence

All Participants

Thirty-four (19.5%) employees had current neck or upper extremity CTD's, as determined by questionnaire and physical exam. Most CTDs involved the hand and wrist. Employees in the HE group were 3.6 times more likely to satisfy the point prevalence case definition than employees in the LE group (95% CI 0.91-14.28). (Table 5)

Women Participants

Twenty-nine (20.4%) employees, all in HE departments, had current neck or upper extremity CTDs, as determined by questionnaire and physical exam. Most CTDs involved the hand and wrist. (Table 6)

2. Medical Management

a. Access to the Plant Nurse

Of the 135 employees having any neck or upper extremity symptom, 31 (23.0%) said that at some point during the past year their foreman/supervisor did not let them leave the line to see the plant nurse. Of the 62 employees satisfying the period prevalence case definition, 13 (21.0%) reported the same thing, as did 10 (29.4%) of the 34 employees satisfying the point prevalence case definition.

b. Physician Evaluation

In the last year, physicians to which the company referred employees treated 19 (14.1%) of the employees having any neck or upper extremity symptom, 8 (12.9%) who satisfied the period prevalence case definition, and 6 (17.7%) who satisfied the point prevalence case definition (as reported on questionnaires by employees).

c. Missed and Restricted Days

Of the employees who satisfied the period prevalence case definition, 9 (14.5%) were given work days off to recover from their injuries, and 20 (32.3%) were given light or restricted jobs. Of the workers who satisfied the point prevalence case definition, 4 (11.8%) were given work days off and 12 (35.3%) were given light or restricted jobs.

3. CTD Prevention Program

a. Job Rotation

Forty-two (29.6%) employees in HE departments said that they were involved in a job rotation program. The mean number of days/week rotated was 2.1, the mean number of hours/day rotated 6, and the mean number of jobs rotated to 2. The jobs most commonly rotated to were in HE departments (such as cut up and deboning). Some employees reported that they were rotated to fill vacancies on the production line, rather than to reduce ergonomic stress.

b. Provision of Sharp Knives and Scissors

Of the employees in HE departments who used knives, 18 (50.0 %) said that they received them once/day, 16 (44.4%) said twice/day, and 2 (5.6%) said more than three times/day. Sixty-six and seven-tenths percent (66.7%) said that they did not receive newly sharpened knives often enough. Many employees reported that when they did receive newly sharpened knives, the tools were, in reality, not very sharp.

Of the employees in HE departments who used scissors, 40 (83.3%) said that they received them once/day, 7 (14.6%) said twice/day, and 1 (2.1%) said three times/day. Sixty-seven and nine-tenths percent (67.9%) said that they did not receive newly sharpened scissors often enough.

c. Training

Nine (6.5%) employees in HE departments were trained to recognize symptoms of carpal tunnel syndrome or tendonitis when they first started working.

Sixty-nine (68.3%) employees in HE departments who cut or sliced meat were trained to do so when they first started working at the plant. Most were trained for 4 weeks, 8 hours/day. Fifty-seven (82.7%) said that this was enough time.

Ten (13.2%) employees in HE departments who used knives were trained to sharpen them when they first started working.

Health Hazard Evaluation Report No. 89-307

(The statistics cited in this section may not reflect recent changes in the CTD prevention program.)

B. Robersonville

1. Prevalence Rates

   a. Participation

   All Participants

   One hundred twenty (120) employees were interviewed and examined. (None who were asked to participate refused.) Of these, 69.2% worked in higher exposure (HE) departments (receiving-5.0%, evisceration-18.3%, grading-6.7%, [whole bird] packing-2.5%, cut up-8.3%, table deboning-14.2%, and cone deboning-14.2%) and 30.8% worked in lower exposure (LE) departments (maintenance-6.7%, sanitation-8.3%, quality assurance-7.5%, and clerical-8.3%). (Table 7)

   Women Participants

   Eighty-four (84) employees were women. Of these, 81.0% worked in HE departments (evisceration-19.0%, grading-7.1%, [whole bird] packing-3.6%, cut up-12.0%, table deboning-19.0%, and cone deboning-20.2% [0.1% rounding error]), and 19.0% worked in LE departments (quality assurance-7.1% and clerical-11.9%). (0.1% rounding error) (Table 8)

   b. Demographics

   All Participants

   Seventy percent (70.0%) of participants were female and 30.0% were male. Eighty-five and eight-tenths percent (85.8%) were black and 14.2% were white. The mean age was 33 years, the mean length of employment 4.1 years, and the mean "time at current job" 39.0 months. The HE group had a greater percentage of women and blacks than the LE group. (Table 9)

   Women Participants

   Eighty-nine and three-tenths (89.3%) of women were black and 10.7% were white. The mean age was 34 years, the mean length of employment 4.1 years, and the mean "time at current job" 37.7 months. The HE group had a greater percentage of blacks than the LE group. (Table 10)

Health Hazard Evaluation Report No. 89-307

c. Period Prevalence

All Participants

Eighty-eight (73.3%) employees reported symptoms compatible with neck or upper extremity CTDs over the one-year study period and 24 (20.0%) satisfied the period prevalence case definition. Employees in the HE group were 2.6 times more likely to have symptoms (95% CI 1.65-4.01) and 10.1 times more likely to satisfy the period prevalence case definition 95% CI 1.42-72.24) than employees in the LE group. (Table 11)

Women Participants

Seventy (83.3%) employees reported symptoms compatible with neck or upper extremity CTDs over the one-year study period and 19 (22.6%) met the period prevalence case definition. Employees in the HE group were 2.5 times more likely to have symptoms (95% CI 1.33-4.74) and 4.2 times more likely to satisfy the period prevalence case definition (95% CI 0.6-29.01) than employees in the LE group. (Table 12)

d. Point Prevalence

All Participants

Nine (7.5%) employees, all in the HE group, had current neck or upper extremity CTDs, as determined by questionnaire and physical exam. Most CTDs involved the hand and wrist. (Table 11)

Women Participants

Seven (8.3%) employees, all in the HE group, had current neck or upper extremity CTDs, as determined by questionnaire and physical exam. Most CTDs involved the hand and wrist. (Table 12)

2. Medical Management

a. Access to the Plant Nurse

Of the 88 employees having any neck or upper extremity symptom, 5 (5.7%) said that at some point during the past year their foreman/supervisor did not let them leave the line to see the plant nurse. Of the 24 employees satisfying the period prevalence case definition, 2 (8.3%) reported the same thing, as did 9 (11.1%) of the employees satisfying the point prevalence case definition.

Health Hazard Evaluation Report No. 89-307

b. Physician Evaluation

In the last year, physicians to which the company referred employees treated 5 (5.7%) of the employees having any neck or upper extremity symptom, 2 (8.3%) who satisfied the period prevalence case definition, and 2 (22.2%) who satisfied the point prevalence case definition (as reported by employees).

c. Missed and Restricted Days

None of the employees who satisfied the period prevalence case definition and none who satisfied the point prevalence case definition were given workdays off to recover from their injuries. Six (25.0%) employees from the former group and 4 (44.4%) from the latter group were given light or restricted jobs.

3. CTD Prevention Program

a. Job Rotation

Twenty-two (26.8%) employees in HE departments said that they were involved in a job rotation program. The mean number of days/week rotated was 2, the mean number of hours/day rotated 5, and the mean number of jobs rotated to 3. The jobs rotated to were in HE departments (such as evisceration, grading, whole bird packing, cut up, table deboning and cone deboning). Some employees reported that they were rotated to fill vacancies on the production line, rather than to reduce ergonomic stress.

b. Provision of Sharp Knives and Scissors

Of the employees in HE departments who used knives, 3 (15.0%) said that they received them once/day, 13 (65.0%) said twice/day, 1 (5.0%) said three times/day, and 3 (15.0%) said more than three times/day. Forty-four and four-tenths percent (44.4%) said that they did not receive newly sharpened knives often enough. Many employees reported that when they did receive newly sharpened knives, the tools were, in reality, not very sharp.

Of the employees in HE departments who used scissors, 17 (47.2%) said that they received them once/day, 15 (40.5%) said twice/day, and 5 (13.5%) said three times/day. Sixty-seven and six-tenths percent (67.6%) said that they did not receive newly sharpened scissors often enough.

c. Training

Two (2.8%) employees in HE departments were trained to recognized symptoms of carpal tunnel syndrome or tendonitis when they first started working.

Forty-five (86.5%) employees in HE departments who cut or slice meat were trained to do so when they first started working at the plant. Most were trained for 4 weeks, 7-8 hours/day. Thirty-three (75.0%) said that this was enough time.

Thirteen (52.0%) employees in HE departments who used knives were trained to sharpen them when they first started working.

(The statistics cited in this section may not reflect recent changes in the CTD prevention program.)

VI. DISCUSSION

A. Lewiston

1. Prevalence Rates

For all employees who participated in this investigation, the overall period prevalence rate was 36%. The period prevalence rate for hand and wrist CTDs was 25%. The respective point prevalence rates were 20% and 17%. All of these rates were considerably higher in higher exposure (HE) departments than in lower exposure (LE) departments. As determined by questionnaire, employees in HE departments were 4.4 times more likely to develop CTDs than employees in LE departments. As determined by questionnaire and physical examination, they were 3.6 times more likely to develop CTDs. (When only women were considered, employees in HE departments also had higher prevalence rates than employees in LE departments.) The markedly elevated CTD rates were not due to misclassification of exposure or disease, survivor bias, or confounding factors.

a. Exposure Misclassification

One employee in a LE department who satisfied both the period and point prevalence case definitions said that his CTD began during a previous job, when he was working in a HE department. As a result of this misclassification, (1) period and point prevalence rates may have been slightly underestimated in HE departments and slightly overestimated in LE departments, and (2) relative rates may have been slightly underestimated. (No employees currently in HE departments said that their CTDs started when they were working in LE departments.)

In addition, since a detailed ergonomic analysis was not done on every job, it is possible that some jobs in LE departments may have had significant risk factors for neck and upper extremity CTDs.

b. Disease Misclassification

To minimize disease misclassification, standardized epidemiologic techniques that have been employed in other studies were used.[1]

Random disease misclassification may have occurred in determining point prevalence rates, due to considerable variation among physicians in terms of the frequency with which certain findings were noted on physical examination. However, because the diagnostic criteria used in this investigation were more stringent than those that have been used in similar NIOSH investigations, disease misclassification is unlikely to be responsible for the high point prevalence rates observed. In fact, the stringent criteria may have actually resulted in an underestimation of point prevalence rates in both HE and LE departments.

c. Survivor Bias

There is considerable annual employee turnover, suggesting that survivor bias may be a substantial problem. "Survivors" (i.e. people who are working) are usually healthier (that is, lacking illnesses or injuries that would interfere with work) than those people who leave employment. The "survivor effect" has been described in studies of other industries and is an inherent bias in the cross-sectional design of this investigation.[2] Survivor bias is nonrandom, affecting HE departments more than LE departments. Since this is likely to decrease the prevalence rates for HE departments, such a bias could have led to an underestimation of relative rates.

d. Potential Confounders

Several factors were considered as potential confounders of the association between the occurrence of CTDs and work in HE departments. These factors included age, race, and gender.

Age

Participants who were 30 years of age or younger were slightly more likely than older workers to be assigned to HE departments (Relative Rate 1.16, 95% CI 1.00 - 1.34). As a result, it was not surprising that workers 30 years of age or younger may have had a slightly increased rate of developing CTDs than older workers (Relative Period Prevalence Rate 1.11, 95% CI 0.74 - 1.66; Relative Point Prevalence Rate 1.40, 95% CI 0.75 - 2.58). However, age did not affect the association between the occurrence of CTDs and work in HE departments. (Participants who were 30 years of age or younger had period prevalence and point prevalence rates similar to those of older workers, when separately comparing HE departments and LE departments.)

Race

Black participants were far more likely to work in HE departments than in LE departments. In fact 99.3% of HE participants were black, compared to 43.8% of LE participants. Because of the high percentage of blacks in HE departments, the association between the occurrence of CTDs and race could not be evaluated. It should be pointed out, however, that an association between race and CTDs has not been well described in the literature. It is unlikely, therefore, that race confounds the association between the occurrence of CTDs and work in HE departments.

Gender

Women participants were 1.37 times (95% CI 1.04 - 1.81) more likely than men to be assigned to HE departments. Independent of this, women may also have had higher period prevalence (1.52, 95% CI 0.81 - 2.87) and point prevalence rates (1.31, 95% CI 0.55 - 3.11) than men for CTDs. (Whereas men in HE departments had a period prevalence rate 1.80 times that of other men, women in HE departments had a period prevalence rate 8.69 times that of other women.) For this reason, an analysis stratifying on sex was done and the results for women noted in this report. (The results for men were not noted, as the sample sizes were quite small.)

It is important to note, however, that since (1) all jobs in HE departments probably do not carry equal risk for CTDs, and (2) the number of men and women performing identical jobs is small, it is impossible to definitively determine if women really have higher CTD rates than men.

2. **Medical Management and CTD Prevention Program**

It is difficult to evaluate the Lewiston facility's management of injured workers and its injury prevention program based on information compiled from the questionnaire that was administered to employees. In order to examine medical management and CTD prevention, more information than was collected in NIOSH's survey is needed. Nevertheless, some findings deserve comment.

One sound ergonomic principle is that the employee is the best judge of the effectiveness of his/her tools. Accordingly, the low percentage of employees who thought that they were receiving adequate numbers of sharp knives and scissors each day is troublesome.

Health Hazard Evaluation Report No. 89-307

Equally troublesome is the fact that at least some jobs used in the current rotation scheme are inappropriate. Rotation between two HE jobs that both stress the neck and upper extremity is likely to be only modestly effective (or even ineffective) in the prevention and early treatment of CTDs.

Finally, according to questionnaire responses, employees at Lewiston having any neck or upper extremity symptoms were 4.3 times more likely at some point in the past year to have been denied access to the plant nurse by their foreman/supervisor than employees at Robersonville (95% CI 1.71-10.68).

B. <u>Robersonville</u>

1. <u>Prevalence Rates</u>

For all employees who participated in this investigation, the overall period prevalence rate was 20%. The period prevalence rate for hand and wrist CTDs was 16%. The respective point prevalence rates were 8% and 7%. All of these rates were considerably higher in higher exposure (HE) departments than in lower exposure (LE) departments. As determined by questionnaire, employees in HE departments were 10.1 times more likely to develop CTDs than employees in LE departments. (When only women were considered, employees in HE departments also had higher prevalence rates than employees in LE departments.) The markedly elevated CTD rates were not due to misclassification of exposure or disease, survivor bias, or confounding factors.

a. <u>Exposure Misclassification</u>

One employee in a LE department who satisfied the period prevalence case definition said that her CTD began during a previous job, when she was working in a HE department. As a result of this misclassification, (1) period and point prevalence rates may have been slightly underestimated in HE departments and slightly overestimated in LE departments, and (2) relative rates may have been slightly underestimated. (No employees currently in HE departments said that their CTDs started when they were working in LE departments.)

In addition, since a detailed ergonomic analysis was not done on every job, it is possible that some jobs in LE departments may have had significant risk factors for neck and upper extremity CTDs.

b. <u>Disease Misclassification and Survivor Bias</u>

See VI. A. 1. b.

Health Hazard Evaluation Report No. 89-307

c. Potential Confounders

Several factors were considered as potential confounders of the association between occurrence of CTD's and work in HE departments. These included age, race, and gender.

Age

There was very little difference in age between participants who worked in HE and LE departments. Therefore, age was not considered to be a confounding factor

Race

Black participants were far more likely to work in HE departments than in LE departments. In fact 100% of HE participants were black, compared to 54.1% of LE participants. Because of the higher percentage of blacks in HE departments, the association between race and the occurrence of CTDs could not be evaluated. It should be pointed out, however, that an association between race and CTDs has not been well described in the literature. It is unlikely, therefore, that race confounds the association between occurrence of CTDs and work in HE departments.

Gender

Women participants were 1.94 times (95% CI 1.30 - 2.90) more likely than men to work in HE departments. Independent of this, women may also have had higher period prevalence (1.54, 95% CI 0.62 - 3.84) and point prevalence rates (1.50, 95% CI 0.33 - 6.87) than men for CTDs. (Whereas men in HE departments had a period prevalence rate 0.90 times that of other men, women in HE departments had a period prevalence rate 5.32 times that of other women. Similar trends were found for point prevalence rates.) For this reason, an analysis stratifying on sex was done and the results for women noted in this report. (The results for men were not noted, as the sample sizes were quite small.)

It is important to note, however, that since (1) all jobs in HE departments probably do not carry equal risk for CTDs, and (2) the number of men and women performing identical jobs is small, it is impossible to definitively determine if women really have higher CTD rates than men.

Health Hazard Evaluation Report No. 89-307

2. Medical Management

It is difficult to evaluate the Robersonville facility's management of injured employees and its injury prevention program based on information compiled from the questionnaire that was administered to employees. In order to examine medical management and CTD prevention, more information than was collected in NIOSH's survey is needed. Nevertheless, some findings deserve comment.

One sound ergonomic principle is that the employee is the best judge of the effectiveness of his/her tools. Accordingly, the low percentage of employees who thought that they were receiving adequate numbers of sharp knives and scissors each day is troublesome.

Equally troublesome is the fact that at least some jobs used in the current rotation scheme are inappropriate. Rotation between two HE jobs that both stress the neck and upper extremity is likely to be only modestly effective (or even ineffective) in the prevention and early treatment of CTDs.

Finally, certain parts of the Perdue, Inc. "Repetitive Motion Disorders Action Plan" for Robersonville (dated 11-15-88) deserve comment. This document states that "all new hires shall be advised in orientation with Human Resources and during the physical, with the nurses, that the company will recommend them to follow a strict preventative program around-the-clock during the term of their probationary period." The program includes taking 2 Ibuprofen tablets (400 mgs.) four times per day, 100 mgs. Vitamin B6 daily, 1600 units Vitamin E daily, and 4 grams Vitamin C daily. "Any team member who refuses to follow preventative medical treatment will be first required to furnish Perdue with a note from their personnel physician stating that medications used by the Industrial Nurse can not be taken by the team member for valid medical reasons or the team member may be allowed to sign a waiver refusing the recommended medication after counselling by the Industrial Nurse under which he/she assumes the responsibility of any ROM problems during the [probationary] period."

Although vitamins, anti-inflammation medications, and a variety of exercise programs have been advocated as methods of preventing work-related CTDs of the upper extremity, NIOSH investigators are unaware of any valid, scientific research that establishes the effectiveness of these interventions.[3,4] In addition, these interventions are not adequate substitutes for the effective engineering and administrative control of CTDs. Finally, the regular consumption of therapeutic amounts of ibuprofen is associated with a risk of various adverse health effects, including perinatal complications.[5]

Health Hazard Evaluation Report No. 89-307

C. Overall Remarks

CTDs are a serious problem in the meat packing and poultry processing industries because of the repetitive and forceful nature of jobs in these sectors. CTDs have been reported in chicken, turkey, pig, and cattle slaughterhouses in the United States and abroad.[2,6-12] The markedly elevated period and point prevalence rates seen in HE departments at the Lewiston and Robersonville facilities is further evidence of this problem.

It is unclear why prevalence rates were lower at the Robersonville facility than at the Lewiston facility. The difference cannot be attributed to the prophylactic use of vitamins and anti-inflammation medications at Robersonville, as only 11 of the 120 employees from this plant who participated in the investigation reported having received such prophylaxis. (Most participants began working before Perdue Farms, Inc. instituted its "Repetitive Motion Disorder Plan.")

VII. RECOMMENDATIONS

The prevention and management of work-related upper extremity CTDs can be divided into 3 areas: engineering, administrative, and medical controls.

A. Engineering Controls

All jobs that (1) are in higher exposure (HE) departments, or (2) have known risk factors for CTDs (i.e. high repetition, forceful exertion, and/or extreme or awkward posture) should be carefully assessed ergonomically to determine the need for job redesign.

After an ergonomic assessment is made, the following recommendations (if appropriate) should be implemented to reduce CTD risk factors.

1. Highly repetitive movements can be reduced in frequency by either slowing down the main conveyor or providing diverging conveyors off the main one so that tasks can be performed at slower rates. Restructuring jobs so that employees' tasks are varied, increasing the number of employees, and automating processes can also reduce repetitiveness. (At Lewiston and Robersonville, work areas are already cramped. Therefore, increasing the numbers of employees without increasing the size of work areas is not recommended, as this could result in an increase in traumatic injuries such as lacerations and amputations.)

2. Excessively forceful exertions can be reduced in intensity by using mechanical devices to aid in deboning and cutting. Maintaining sharp cutting edges on knives and scissors and automating processes can also reduce excessively forceful exertions.

3. Extreme postures can be eliminated by means such as providing work stations that accommodate the height and reach limitations of different size employees.

The design of effective engineering controls is best done with input from employees and supervisors who will be affected by changes.

B. Administrative Controls

Training, job rotation, and rest pauses are recommended as administrative control measures. These methods, however, have not been validated in scientific studies. None of them are as important as engineering controls.

1. Training

New employees should see appropriate demonstrations and be given time to practice proper cutting techniques and knife care. They should also be given the opportunity to condition their muscles and tendons prior to working at full capacity. Conditioning can be accomplished by putting new employees in slower paced lines, varying each employee's tasks, and rotating each employee through different jobs. Training should be done over the course of several weeks.

2. Job Rotation

The aim of job rotation is to alleviate the physical fatigue and stress of particular muscle-tendon-nerve groups by rotating employees among jobs that require the use of different muscles and tendons. Caution, however, must be used in deciding which jobs to rotate employees through. Although different jobs may appear to require the use of different muscle-tendon-nerve groups, they may actually stress the same area. In addition, rotation schedules should be designed to ensure that the benefits derived by some employees are not compromised by subjecting other employees (who must share the ergonomically hazardous tasks) to excessive musculoskeletal stress.

3. Rest Pauses

Rest pauses are needed to relieve fatigued muscles and tendons.

C. Medical Controls

1. Health Care Providers

Health care providers should be knowledgeable in (1) the prevention, recognition, treatment, and rehabilitation of CTDs, and (2) the basic principles of ergonomics and epidemiology. In addition, they should be familiar with OSHA recordkeeping requirements. At the minimum, an occupational health nurse should be available on each shift.

Health Hazard Evaluation Report No. 89-307

2. **Workplace Walk-Throughs**

   Health care providers should conduct routine, systematic workplace walk-throughs to understand processes and work practices, identify CTD risk factors, and become aware of any potential light duty jobs. Walk-through surveys should be conducted every month or whenever processes and work practices change significantly.

3. **Catalog of Job Descriptions**

   An ergonomist or other similarly qualified person should ergonomically assess every job and provide health care personnel with the results of this assessment.

4. **Active CTD Surveillance**

   a. **Survey of Symptoms and CTDs**

   The first goal of the surveillance program is to determine (1) the types of symptoms and CTDs that are occurring, and (2) whether the incidence of these problems is increasing, decreasing or remaining the same.

   To accomplish this goal, the OSHA 200 Log should be analyzed and a questionnaire administered to all workers. The questionnaire should be administered once a year. They should elicit information regarding (1) the location, frequency, and duration of work-related CTD symptoms, and (2) employees' perceptions about causes of these problems. Employees' names should not be required on questionnaires, as fear of repercussions for reporting symptoms can lead to the collection of inaccurate data. (A sample questionnaire is provided in Appendix A.)

   b. **Health Survey**

   The second goal of the surveillance program is to detect CTDs in order to facilitate early treatment.

   To accomplish this goal, a questionnaire (such as the one used at the Lewiston and Robersonville facilities) should be administered to all employees and a brief physical examination performed by a health care provider. This should be done once a year or after an employee changes jobs. It is important to note that this is not a preplacement exam and should not be used to screen out workers.

5. **Employee Education.**

   All employees, including supervisors and other management personnel, should be educated about the prevention, recognition, treatment, and rehabilitation of CTDs. The

Health Hazard Evaluation Report No. 89-307

information should be reinforced by health care providers during workplace walk-throughs and physical examinations. New employees should be educated during orientation. Education programs facilitate the early recognition of CTDs (prior to the development of severe, disabling conditions), and increase the likelihood of compliance with prevention and treatment programs.

6. <u>Prophylactic Use of Vitamins and Anti-Inflammation Medications</u>

    Although vitamins, anti-inflammation medications, and a variety of exercise programs have been advocated as effective methods of preventing work-related CTDs of the upper extremity, NIOSH is unaware of any valid, scientific research that establishes the effectiveness of these interventions.[3,4] In addition, these interventions are not adequate substitutes for effective engineering and administrative controls. Finally, the regular consumption of therapeutic amounts of ibuprofen, a commonly used anti-inflammatory agent, is associated with a risk of various adverse health effects, including perinatal complications.[5]

7. <u>Evaluation, Treatment, and Follow-up of CTDs</u>

    If CTDs are recognized and treated early in their development, debilitating conditions may be prevented. Symptomatic employees should be guaranteed access to the plant nurse. The nurse, in turn, should take a medical history and perform a limited musculoskeletal physical examination. The examination should include inspection, palpation, an assessment of strength and range of motion (passive, active, and resisted), and the performance of various diagnostic maneuvers (such as Tinel's test, Phalen's test, and Finkelstein's test). Laboratory tests, X-rays and other diagnostic procedures should not be done routinely at this stage.

    Any employee with (1) numbness or crepitus, (2) a positive Tinel's, Phalen's, or Finkelstein's test, or (3) evidence of medial or lateral epicondylitis or a rotator cuff injury should be referred to a physician. If a physician referral is not necessary, the treatment regime outlined in the Upper Extremity Cumulative Trauma Disorder Algorithm (Appendix B) should be followed. This algorithm outlines a conservative approach to treating CTD's, employing the use of the following therapies:

    a. <u>Non-Steroidal Anti-Inflammation Medications</u>

        These agents are helpful in reducing inflammation and pain.

    b. <u>Ice</u>

        Ice reduces inflammation and should be used even if no overt signs of inflammation (i.e. redness, warmth, or swelling) are present.

Ice should be applied to affected areas 4 times per day, for 20 minutes each time. Heat treatments should be used only for muscle strains.

c. Exercises

Once CTDs have occurred, in general, passive range of motion exercises should be initiated. (If active exercises are used they should be administered under the supervision of an occupational health nurse, a physician, or a physical therapist. If they are performed improperly, they can aggravate existing conditions.)

d. Light/Restricted Duty

Job reassignment must be done with knowledge of whether new tasks will require use of injured muscles or tendons, or put pressure on injured nerves. Inappropriate reassignment can exacerbate CTDs and result in permanent disability.

e. Splints

Splints are helpful in immobilizing symptomatic muscles, tendons, and nerves. They should not be used during work unless the employee has been transferred from his/her job and it has been determined by the health care provider that the new job does not stress the muscle-tendon-nerve group being splinted.

It is important to note that although the algorithm includes many commonly accepted treatments for CTDs, the effectiveness of these treatments has not been validated in scientific studies. In addition, many of the treatments can have serious side effects: anti-inflammation medications can cause gastrointestinal irritation and bleeding; and active stretching exercises can exacerbate CTD symptoms; THEREFORE, ANY CTD PREVENTION PROGRAM SHOULD PLACE PRIMARY EMPHASIS ON REMOVING CTD RISK FACTORS, RATHER THAN RELYING ON THE MEDICAL TREATMENT OF SYMPTOMATIC EMPLOYEES.

Concerning surgery, (1) "second opinions" should be obtained before surgery is done, and (2) after surgery, appropriate time off work should be provided to allow all injured muscle-tendon-nerve groups and operative sites to heal. The exact number of days off work will vary from employee-to-employee. For carpal tunnel surgery, the following averages have been proposed as guidelines:[13,14]

When returning to a nonrepetitive, non-forceful job (job with cycle time of 5 minutes or more; that never requires lifting objects over 1 pound, using hand tools, or pinching or gripping) - <u>3 weeks off</u> (minimum 10 days),

When returning to a low-moderately repetitive, low-moderately forceful job (job with cycle time between 30 seconds and 5 minutes; that requires lifting objects less than 2 pounds during most job cycles or occasionally using hand tools) - <u>6 weeks off</u> (minimum 21 days),

When returning to a highly repetitive, highly forceful job (job with cycle time less than 30 seconds; that requires lifting more than 2 pounds during most job cycles or regularly using hand tools requiring forceful exertions) - <u>12 weeks off</u> (minimum 42 days),

It must be emphasized that these are averages. Some workers may require more time off or less time off, depending on individual responses to surgery. In addition, these recovery times are the opinions of recognized experts or authors of published articles and do not represent NIOSH policy. These experts and authors emphasize that recovery time is generally a matter of 2-3 months, and <u>not</u> 2-3 weeks.

After an employee has been away from work for medical reasons, he/she should be evaluated by a physician. This evaluation should include an assessment of his/her work capabilities. The physician should either view the employee's job on videotape or, preferably, see it first-hand in the plant.

Every time an employee is seen by a health care provider, the encounter should be documented in the employee's medical records.

D. Other

The CTD prevention and management program should be developed and implemented with input from health care providers, management, and employees. To accomplish this, a committee composed of representatives of these groups should be set up, with members of the committee (1) being educated about the basic ergonomic and medical principles of CTDs, (2) overseeing an ergonomic assessment of the workplace, (3) designing a CTD prevention and management program, and (4) evaluating the effectiveness of the program.

VIII. REFERENCES

1. Silverstein BA, Fine LJ, Armstrong TJ. Hand wrist cumulative trauma disorders in industry. <u>British Journal of Industrial Medicine</u>. 1986; 43:779-784.

2. Viikari-Juntura E. Neck and upper limb disorders among slaughterhouse workers. <u>Scandinavian Journal of Work and Environmental Health</u>. 1983; 9:283-290.

3. Amadio PC. Carpal tunnel syndrome, pyridoxine and workplace. <u>Journal of Hand Surgery</u>. 1987; 2a(part 2):875-879.

4. Silverstein BA, Armstrong JA, Longmate A, Woody D. Can in-plant exercises control musculoskeletal symptoms? *Journal of Occupational Medicine*. 1988; 38:922-927.

5. Kelley WN, Harris ED, Ruddy S, Sledge, CB. *Textbook Of Rheumatology*. W.B. Saunders, 1981. page 750.

6. Armstrong TJ, Foulke JA, Joseph BS, Goldstein SA. Investigation of cumulative trauma disorders in a poultry processing plant. *American Industrial Hygiene Journal*. 1982; 43:103-116.

7. Falck B, Aarnio P. Left-sided carpal tunnel syndrome in butchers. *Scandinavian Journal of Work and Environmental Health*. 1983; 9:291-297.

8. Finkel ML. The effects of repeated mechanical trauma in the meat industry. *American Journal of Industrial Medicine*. 1985; 8:375-379.

9. Masear VR, Hayes JM, Hyde AG. An industrial cause of carpal tunnel syndrome. *The Journal of Hand Surgery*. 1986; 11a:222-227.

10. Muffly-Elsey D, Flinn-Wagner S. Proposed screening tool for the detection of cumulative trauma disorders of the upper extremity. *The Journal of Hand Surgery*. 1987; 12a:931-935.

11. Hales T, Habes D, Fine L, Hornung R, Boiano J. *National Institute of Occupational Safety and Health (NIOSH) Health Hazard Evaluation Report 88-180-1958 (John Morrell and Company)*. 1989.

12. Hales T, Fine L. *National Institute of Occupational Safety and Health (NIOSH) Health Hazard Evaluation Report 89-251-1997 (Cargill)* 1989.

13. Sidney Blair. Chairman, Department of Orthopedics and Rehabilitation, Loyola University Medical Center, Chicago, Illinois, personal communication.

14. Dawson DM, Hallett M, Millender L. *Entrapment Neuropathies*. Little, Brown and Co. 1983. page 59.

IX. AUTHORSHIP AND ACKNOWLEDGEMENTS

Report prepared by:
    Stuart Kiken, M.D., M.P.H.
    William Stringer, M.S.
    Lawrence Fine, M.D., Dr.P.H.
    Thomas Sinks, Ph.D.
    Shiro Tanaka, M.D., M.S.

Originating office:
    Hazard Evaluations and Technical Assistance Branch
    Division of Surveillance, Hazard Evaluations, and Field Studies

Health Hazard Evaluation Report No. 89-307

The authors would like to acknowledge the following NIOSH personnel: Tom Hales, for designing the protocol and assisting in report preparation; Kathy Watkins, for preparing the questionnaire; B.J. Haussler, for overseeing the administration of questionnaires; Calvin Cook and Chris Hollis, for administering questionnaires; Ray Aldefer and Anthony Suruda, for doing physical examinations; and Don Bates, for supervising data entry.

In addition, the authors would like to thank the following North Carolina Department of Labor personnel: James Oppold, Kevin Hanley, David Lipton, and Mary Carol Lewis, for facilitating the investigations; and Mark Jackson, S.B. White, Jean Williams, and Bion Brewer, for administering questionnaires.

Finally, the investigations would not have possible without the cooperation of the management and employees of Perdue Farms, Inc.

X. DISTRIBUTION AND AVAILABILITY

Copies of this report are temporarily available upon request from NIOSH, Hazard Evaluations and Technical Assistance Branch, 4676 Columbia Parkway, Cincinnati, Ohio 45226. After 90 days, the report will be available through the National Technical Information Service (NTIS), 5285 Port Royal, Springfield, Virginia 22161. Information regarding its availability through NTIS can be obtained from the NIOSH Publications Office at the Cincinnati address. Copies of this report have been sent to:

1. The North Carolina Department of Labor, Division of Occupational Safety and Health.
2. Perdue Farms, Inc.

For the purpose of informing affected employees, copies of this report should be posted by Perdue Farms, Inc. in a prominent place that is accessible to employees for a period of 30 calendar days.

TABLE 1

PARTICIPATION
LEWISTON - ALL EMPLOYEES

PERDUE FARMS, INC.
LEWISTON, NORTH CAROLINA
HETA 89-307

Department	Number of Participants	Total Number of Employees*
Receiving	17	43
Evisceration	17	139
Grading	2	25
Stretch Bag	1	27
Cut Up	40	233
Deboning	65	237
Maintenance	8	104
Quality Control	11	16
Clerical	13	NA**
Total	174	

* day shift (except for one employee, all participants were from the day shift)
** not available

TABLE 2

PARTICIPATION
LEWISTON - WOMEN

PERDUE FARMS, INC.
LEWISTON, NORTH CAROLINA
HETA 89-307

Department	Number of Participants
Evisceration	17
Grading	2
Stretch Bag	1
Cut Up	40
Deboning	62
Quality Control	8
Clerical	12
Total	142

TABLE 3

AGE, RACE, GENDER, AND LENGTH OF EMPLOYMENT BY EXPOSURE GROUP
LEWISTON - ALL PARTICIPANTS

PERDUE FARMS, INC.
LEWISTON, NORTH CAROLINA
HETA 89-307

	Overall	HE*	LE**	95% CI
Age (Years)	31.1	30.4	34.4	
Race				
% Black	89.1%	99.3%	43.8%	2.56-116.53
% White	10.9%	0.7%	56.3%	
Gender				
% Male	18.4%	14.1%	37.5%	1.04-1.81
% Female	81.6%	85.9%	62.5%	
Length of Employment (Years)	5.5	5.4	5.8	

* higher exposure
** lower exposure

TABLE 4

AGE, RACE, AND LENGTH OF EMPLOYMENT BY EXPOSURE GROUP
LEWISTON - WOMEN

PERDUE FARMS, INC.
LEWISTON, NORTH CAROLINA
HETA 89-307

	Overall	HE*	LE**	95% CI
Age (Years)	31.4	30.9	34.0	
Race				
% Black	90.8%	99.2%	40.0%	1.85-80.20
% White	9.2%	0.8%	60.0%	
Length of Employment (Years)	5.4	5.4	5.2	

* higher exposure
** lower exposure

TABLE 5

CTD SYMPTOMS, PERIOD PREVALENCE CASES, AND POINT PREVALENCE CASES
LEWISTON - ALL PARTICIPANTS

PERDUE FARMS, INC.
LEWISTON, NORTH CAROLINA

HETA 89-307

	Number	Any Symptoms # (%)	Period Prevalence Case # (%)	Point Prevalence Case # (%)
Neck				
HE	142	48 (34%)	13 (9%)	6 (4%)
LE	32	5 (16%)	1 (3%)	1 (3%)
Combined	174	53 (31%)	14 (8%)	7 (4%)
RR*		2.2	2.9	1.3
95% CI		0.94-5.00	0.39-21.44	0.17-10.77
Shoulder				
HE	142	65 (46%)	18 (13%)	4 (3%)
LE	32	9 (28%)	1 (3%)	0 (0%)
Combined	174	74 (43%)	19 (11%)	4 (2%)
RR*		1.6	4.0	
95% CI		0.91-2.91	0.56-29.08	
Elbow				
HE	142	32 (23%)	9 (6%)	0 (0%)
LE	32	1 (3%)	0 (0%)	0 (0%)
Combined	174	33 (19%)	9 (5%)	0 (0%)
RR*		7.2		
95% CI		1.02-50.84		
Hand/Wrist				
HE	142	117 (82%)	42 (30%)	27 (19%)
LE	32	6 (19%)	2 (6%)	2 (6%)
Combined	174	123 (71%)	44 (25%)	29 (17%)
RR*		4.4	4.7	3.0
95% CI		2.13-9.08	1.21-18.54	0.76-12.14
Any Area				
HE	142	123 (87%)	59 (42%)	32 (23%)
LE	32	12 (38%)	3 (9%)	2 (6%)
Combined	174	135 (78%)	62 (36%)	34 (20%)
RR*		2.3	4.4	3.6
95% CI		1.47-3.63	1.48-13.24	0.91-14.28

* relative rate

TABLE 6

CTD SYMPTOMS, PERIOD PREVALENCE CASES, AND POINT PREVALENCE CASES
LEWISTON - WOMEN

PERDUE FARMS, INC.
LEWISTON, NORTH CAROLINA

HETA 89-307

	Number	Any Symptoms		Period Prevalence Case		Point Prevalence Case	
		#	(%)	#	(%)	#	(%)
Neck							
HE	122	45	(37%)	13	(11%)	6	(5%)
LE	20	3	(15%)	0	(0%)	0	(0%)
Combined	142	48	(34%)	13	(9%)	6	(4%)
RR*		2.5					
95% CI		0.84-7.16					
Shoulder							
HE	122	61	(50%)	17	(14%)	4	(3%)
LE	20	6	(30%)	1	(5%)	0	(0%)
Combined	142	67	(47%)	18	(13%)	4	(3%)
RR*		1.7		2.8			
95% CI		0.83-3.33		0.39-19.64			
Elbow							
HE	122	29	(24%)	8	(7%)	0	(0%)
LE	20	0	(0%)	0	(0%)	0	(0%)
Combined	142	29	(20%)	8	(6%)	0	(0%)
95% CI							
Hand/Wrist							
HE	122	106	(87%)	37	(30%)	24	(20%)
LE	20	2	(10%)	0	(0%)	0	(0%)
Combined	142	108	(76%)	37	(26%)	24	(17%)
RR*		8.7					
95% CI		2.33-32.41					
Any Area							
HE	122	112	(92%)	53	(43%)	29	(24%)
LE	20	7	(35%)	1	(5%)	0	(0%)
Combined	142	119	(84%)	54	(38%)	29	(20%)
RR*		2.6		8.7			
95% CI		1.44-4.78		1.27-59.33			

* relative rate

TABLE 7

PARTICIPATION
ROBERSONVILLE - ALL EMPLOYEES

PERDUE FARMS, INC.
ROBERSONVILLE, NORTH CAROLINA
HETA 89-307

Department	Number of Participants		Total Number of Employees	
	Day	Night	Day	Night
Receiving	5	1	10	10
Evisceration	19	3	39	39
Grading	8		10	10
Packing	3		12	12
Cut Up	10		35	35
Table Deboning	15	2	25	25
Cone Deboning	17		50	0
Maintenance	6	2	18*	
Sanitation	1	9	26*	
Quality Control	6	3	10*	
Clerical	10		NA**	
Total	100	20		

* day <u>and</u> night shift combined
** not available

TABLE 8

PARTICIPATION
ROBERSONVILLE - WOMEN

PERDUE FARMS, INC.
ROBERSONVILLE, NORTH CAROLINA
HETA 89-307

Department	Number of Participants
Evisceration	16
Grading	6
Packing	3
Cut Up	10
Table Deboning	16
Cone Deboning	17
Quality Control	6
Clerical	10
Total	84

TABLE 9

AGE, RACE, GENDER, AND LENGTH OF EMPLOYMENT BY EXPOSURE GROUP
ROBERSONVILLE - ALL PARTICIPANTS

PERDUE FARMS, INC.
ROBERSONVILLE, NORTH CAROLINA
HETA 89-307

	Overall	HE*	LE**	95% CI
Age (Years)	33.1	32.8	33.7	
Race				
% Black	85.8%	100.0%	54.1%	1.30-2.90
% White	14.2%	0 %	46.0%	
Gender				
% Male	30.0%	18.1%	56.8%	
% Female	70.0%	81.9%	43.2%	
Length of Employment (Years)	4.1	4.0	4.4	

* higher exposure
** lower exposure

TABLE 10

AGE, RACE, AND LENGTH OF EMPLOYMENT BY EXPOSURE GROUP
ROBERSONVILLE - WOMEN

PERDUE FARMS, INC.
ROBERSONVILLE, NORTH CAROLINA
HETA 89-307

	Overall	HE*	LE**	95% CI
Age (Years)	33.6	33.7	33.1	
Race				
% Black	89.3%	100.0%	43.8%	1.52-63.98
% White	10.7%	0 %	56.3%	
Length of Employment (Years)	4.1	4.2	4.1	

* higher exposure
** lower exposure

TABLE 11

CTD SYMPTOMS, PERIOD PREVALENCE CASES, AND POINT PREVALENCE CASES
ROBERSONVILLE - ALL PARTICIPANTS

PERDUE FARMS, INC.
ROBERSONVILLE, NORTH CAROLINA

HETA 89-307

	Number	Any Symptoms # (%)	Period Prevalence Case # (%)	Point Prevalence Case # (%)
Neck				
HE	83	35 (42%)	4 (5%)	1 (1%)
LE	37	4 (11%)	1 (3%)	0 (0%)
Combined	120	39 (33%)	5 (4%)	1 (1%)
RR*		3.9	1.8	
95% CI		1.49-10.18	0.2-15.23	
Shoulder				
HE	83	35 (42%)	6 (7%)	1 (1%)
LE	37	6 (16%)	0 (0%)	0 (0%)
Combined	120	41 (34%)	6 (5%)	1 (1%)
RR*		2.6		
95% CI		1.2-5.64		
Elbow				
HE	83	11 (13%)	0 (0%)	0 (0%)
LE	37	2 (5%)	0 (0%)	0 (0%)
Combined	120	13 (11%)	0 (0%)	0 (0%)
RR*		2.5		
95% CI		0.57-10.52		
Hand/Wrist				
HE	83	69 (83%)	19 (23%)	8 (10%)
LE	37	10 (27%)	0 (0%)	0 (0%)
Combined	120	79 (66%)	19 (16%)	8 (7%)
RR*		3.1		
95% CI		1.8-5.27		
Any Area				
HE	83	75 (90%)	23 (28%)	9 (11%)
LE	37	13 (35%)	1 (3%)	0 (0%)
Combined	120	88 (73%)	24 (20%)	9 (8%)
RR*		2.6	10.1	
95% CI		1.65-4.01	1.42-72.24	

* relative rate

TABLE 12

CTD SYMPTOMS, PERIOD PREVALENCE CASES, AND POINT PREVALENCE CASES
ROBERSONVILLE - WOMEN

PERDUE FARMS, INC.
ROBERSONVILLE, NORTH CAROLINA

HETA 89-307

	Number	Any Symptoms # (%)	Period Prevalence Case # (%)	Point Prevalence Case # (%)
Neck				
HE	68	32 (47%)	4 (4%)	1 (1%)
LE	16	2 (13%)	1 (6%)	0 (0%)
Combined	84	34 (40%)	5 (6%)	1 (1%)
RR*		3.8	0.9	
95% CI		1.00-14.10	0.11-7.75	
Shoulder				
HE	68	32 (47%)	4 (6%)	1 (1%)
LE	16	5 (31%)	0 (0%)	0 (0%)
Combined	84	37 (44%)	4 (5%)	1 (1%)
RR*		1.5		
95% CI		0.70-3.25		
Elbow				
HE	68	10 (15%)	0 (0%)	0 (0%)
LE	16	1 (6%)	0 (0%)	0 (0%)
Combined	84	11 (13%)	0 (0%)	0 (0%)
RR*		2.4		
95% CI		0.32-17.08		
Hand/Wrist				
HE	68	59 (87%)	14 (21%)	6 (9%)
LE	16	5 (31%)	0 (0%)	0 (0%)
Combined	84	64 (76%)	14 (17%)	6 (7%)
RR*		2.8		
95% CI		1.33-5.78		
Any Area				
HE	68	64 (94%)	18 (27%)	7 (10%)
LE	16	6 (38%)	1 (6%)	0 (0%)
Combined	84	70 (83%)	19 (23%)	7 (8%)
RR*		2.5	4.2	
95% CI		1.33-4.74	0.6-29.01	

* relative rate

APPENDIX A

SURVEILLANCE QUESTIONNAIRE

1. Today's Date   /___/___/ - /___/___/ - 19/___/___/
                      (month)   (day)   (year)

2. Current Department _____

3. Current Job   _____

4. During the past year, have you had pain, aching, stiffness, burning, numbness, or tingling in any of the following areas?

   a. Neck: Yes___1 No___2
   b. Shoulder: Yes___1 No___2
   c. Elbow: Yes___1 No___2
   d. Hand/Wrist Yes___1 No___2

5. If you have had pain, aching, stiffness, burning, numbness, or tingling in the neck:

   a. How often have you had this problem?
      Every 6 months___1   Every 2-3 months___2   Once a month___3
      Once a week   ___4   Daily   ___5

   b. How long does each episode last?
      1 day or less ___1   1 day to 1 week ___2   1 week to 1 month___3
      1-3 months   ___4   3 or more months___5

   c. Was the first time you experienced this problem before or after you started working at this plant? Before___1 After___2

   d. Do you think this problem is caused by work? Yes___1 No___2

      e. If yes, by what in particular? _____
      _____

   f. In the past year have you missed any workdays or been on light/restricted duty because of this problem? Yes___1 No___2

6. If you have had pain, aching, stiffness, burning, numbness, or tingling in the shoulder:

   a. How often have you had this problem?
      Every 6 months___1   Every 2-3 months___2   Once a month___3
      Once a week   ___4   Daily   ___5

   b. How long does each episode last?
      1 day or less ___1   1 day to 1 week ___2   1 week to 1 month___3
      1-3 months   ___4   3 or more months___5

c. Was the first time you experienced this problem before or after you started working at this plant? Before___1 After___2

d. Do you think this problem is caused by work? Yes___1 No___2

   e. If yes, by what in particular? _____
   _____

f. In the past year have you missed any workdays or been on light/restricted duty because of this problem? Yes___1 No___2

7. If you have had pain, aching, stiffness, burning, numbness, or tingling in the elbow:

   a. How often have you had this problem?
      Every 6 months___1   Every 2-3 months___2   Once a month___3
      Once a week   ___4   Daily   ___5

   b. How long does each episode last?
      1 day or less ___1   1 day to 1 week ___2   1 week to 1 month___3
      1-3 months   ___4   3 or more months___5

   c. Was the first time you experienced this problem before or after you started working at this plant? Before___1 After___2

   d. Do you think this problem is caused by work? Yes___1 No___2

      e. If yes, by what in particular? _____
      _____

   f. In the past year have you missed any workdays or been on light/restricted duty because of this problem? Yes___1 No___2

8. If you have had pain, aching, stiffness, burning, numbness, or tingling in the hand/wrist:

   a. How often have you had this problem?
      Every 6 months___1   Every 2-3 months___2   Once a month___3
      Once a week   ___4   Daily   ___5

   b. How long does each episode last?
      1 day or less ___1   1 day to 1 week ___2   1 week to 1 month___3
      1-3 months   ___4   3 or more months___5

   c. Was the first time you experienced this problem before or after you started working at this plant? Before___1 After___2

   d. Do you think this problem is caused by work? Yes___1 No___2

      e. If yes, by what in particular? _____
      _____

   f. In the past year have you missed any workdays or been on light/restricted duty because of this problem? Yes___1 No___2

APPENDIX B

UPPER EXTREMITY (UE) CUMULATIVE TRAUMA DISORDERS (CTD) ALGORITHM

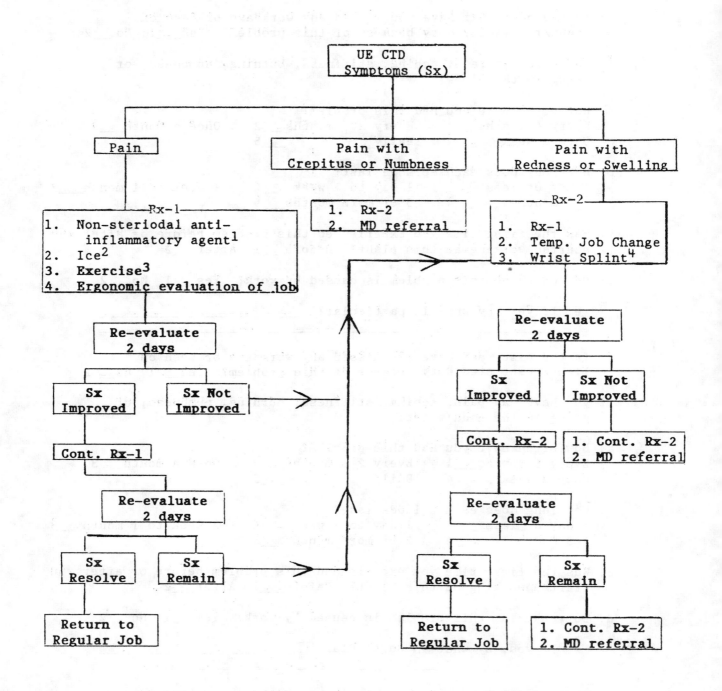

1 - Aspirin 650 mg PO qid or
    Ibuprofen 400 mg PO qid.
2 - Ice to area for 20 minutes qid.
3 - Under nursing supervision for first day.
4 - Only if no wrist bending is required.

HETA 90-251-2128
AUGUST 1991
DOW JONES & COMPANY, INC.
DALLAS, TEXAS

NIOSH INVESTIGATORS:
Deanna Letts, R.N., M.S.
David Nestor, M.S.P.T.
Katharyn A. Grant, Ph.D.
Daniel J. Habes, M.S.E.
Sherry Baron, M.D., M.P.H.
Leslie Copeland, M.S.

I. SUMMARY

On April 27, 1990, the National Institute for Occupational Safety and Health (NIOSH) received a request from the Occupational Safety and Health Administration (OSHA) to conduct a health hazard evaluation at the Dow Jones & Company in Dallas, Texas. The request concerned lower extremity musculoskeletal problems among printers potentially related to poor ergonomic working conditions in the composing room. Specific health effects mentioned in the request included tumors between the metatarsal bones, hammertoes, ingrown nails, bunions, swelling associated with arthritic knees, and heel spurs. In response to this request, a medical and ergonomic investigation was conducted on June 26-27, 1990.

In October/November 1989, several changes occurred in the composing room environment. The carpet was removed and a tiled floor was installed. Stools used by printers during downtime were removed and sitting in the composing room was not allowed unless the employee was assigned a sitting job. In March/April 1990, as a result of management and union negotiations, mats were installed and sit-down breaks during downtime were permitted.

A questionnaire was administered to all 20 printers and 8 of 9 news production staff. Five (25%) of the printers and 1 (13%) news production staff person reported new onset of lower extremity discomfort following the change in composing room floor surface (relative risk [RR]=2.00, 95% confidence intervals [CI]: 0.27, 14.55). Twelve (60%) of the printers and 2 (25%) of the news production staff reported lower extremity discomfort (either new onset or aggravation of a pre-existing condition) following the change of floor (RR=2.40, 95% CI: 0.69, 8.40). Finally, 11 (55%) printers and 1 (13%) news production staff person reported current lower extremity discomfort (either new onset or aggravation of a pre-existing condition) (RR=4.40, 95% CI: 0.67, 28.71). Although the differences in prevalence of lower extremity musculoskeletal problems between the printers and news production staff were not statistically significant (possibly because of the small number of people in the study), more lower extremity problems were consistently reported by the printers than the news staff.

An ergonomic evaluation of the composing room was performed by analysis of photographs and videotaped work tasks. Static standing postures (which, when frequent or sustained, can be stressful and fatiguing to the musculoskeletal system) were observed among printers working at the make-up banks. Make-up banks did not accommodate the short-statured individual as evidenced by employees standing on tiptoe.

> On the basis of the medical and ergonomic investigation, the NIOSH investigator conclude that an ergonomic hazard exists in the composing room among printers due to static standing postures. In addition, unadjustable workstations create an ergonomic hazard for short-statured workers. Recommendations are made in Section IX for modifying the work environment

KEYWORDS: SIC 2711 (Newspapers: Publishing, or Publishing and Printing), journeymen printers, composing room, standing workplace, lower extremity disorders, ergonomics

Health Hazard Evaluation Report No. 90-251

II. INTRODUCTION

On April 27, 1990, the National Institute for Occupational Safety and Health (NIOSH) received a request from the Occupational Safety and Health Administration (OSHA) to conduct a health hazard evaluation at the Dow Jones & Company in Dallas, Texas. The request concerned lower extremity musculoskeletal problems among printers potentially related to poor ergonomic working conditions in the composing room. Specific health effects mentioned in the request included tumors between the metatarsal bones, hammertoes, ingrown nails, bunions, swelling associated with arthritic knees, and heel spurs.

NIOSH investigators conducted a site visit on June 26-27, 1990. The site visit consisted of an opening conference, a walk-through survey of the composing room, medical and ergonomic evaluations, and a closing conference.

III. BACKGROUND

The Dow Jones & Company began production of the Wall Street Journal (WSJ) in Dallas, Texas, in 1948 and moved to its current location in 1965. Two editions (a 2-star edition and a 3-star edition) of the WSJ for the Southwest United States are printed at the Dallas site, with a total production of approximately 95,000 copies per day. In 1972, the facility converted from hot metal printing to cold type. In 1974, approximately 50% of the data to compose the type were being entered at the facility by keyboard operators. During the period between 1985-87 the composing of type at this facility was phased out, and the approximately 12 keyboard operators (all female) were transferred to the make-up area to work as printers. Presently, all type comes directly from computers in New York City (NYC) and is received by typesetting machines in the composing room.

Once the type is received, it is photocopied, trimmed, waxed, and placed on a galley. The galleys holding the type are placed at the ready bank, where they are picked up by the printers and taken to the make-up banks to be used to compose a page. Each composite page, weighing approximately one pound, contains various combinations of news stories, photographs, and advertisements. After a composite page is assembled, a negative is made of it. The negative is sent to the plate-making department, and plates are subsequently sent to the printing department.

The composing room includes 11 make-up banks for composing newspaper pages, the dump area where type is trimmed and waxed, ready banks where type ready for paste-up is placed, typesetter machines driven by computers in NYC, 2 cameras, and advertising and news production areas. There is also a small backroom used for photographic processing.

Health Hazard Evaluation Report No. 90-251

In October/November 1989, several changes occurred in the composing room environment. The carpet was removed and a tiled floor was installed. Stools, used by printers during downtime since approximately 1974, were removed. Subsequently, printers were prohibited from sitting in the composing room (unless they were assigned to a job, such as doing corrections on video display terminals in the advertising department, that required a sitting posture). Reportedly, the only time they were allowed to sit during the workshift was during their lunch time (a half hour break). The union filed a grievance regarding the need for mats, stools, and breaks. An agreement on these issues was reached between union and management representatives in March/April 1990. Subsequently, cushioned mats were installed at the make-up areas and other areas where static standing occurs. Sit-down breaks in the lunch room during downtime were also authorized.

IV. <u>JOB DESCRIPTIONS</u>

Printers are responsible for composing newspaper pages, making corrections, and performing photographic processing. There are 16 full-time and 4 substitute (fill-in for full-time employees) printers. Most of the printers are assigned to make up approximately 2-5 pages each.

News production staff (news staff) are responsible for proofreading, directing corrections, and supervising page make-up. There are 9 news production staff working in the composing room. Reportedly, news staff have more control over their work-rest schedule than printers.

Printers and news staff generally work five 7-hour shifts, Sunday through Thursday; a few work on Friday typesetting classified ads. The majority work the 3:30pm to 11:00pm shift.

V. <u>METHODS</u>

<u>Medical</u>

A NIOSH investigator administered to all 20 of the printers and 8 of the 9 news staff, a questionnaire that focused on lower extremity (LE) musculoskeletal problems potentially related to ergonomic hazards in the composing room. Seventeen of the 20 printers completed the questionnaire in person at the time of the site visit; the other 3 printers and all 8 of the news staff were interviewed by telephone at a later date.

Rates of LE discomfort among printers were compared to news staff. Both printers and news staff worked in the composing room before and after the composing room environment was changed.

Health Hazard Evaluation Report No. 90-251

Ergonomic

The ergonomic evaluation included evaluation of a videotape and photographs, taken by NIOSH personnel, of employees working in the composing room. The videotape and photographs of composing room employees taken by the OSHA investigator were also obtained and reviewed. Finally, a make-up bank was measured.

VI. EVALUATION CRITERIA

In many industries, workers are required to work while standing and walking. Lower extremity musculoskeletal disorders from forced long-term standing and walking has not been well researched in the occupational literature. According to Redfern and Chaffin, in 1983 the American Podiatric Association reported that 83 percent of industrial workers had foot or lower leg problems such as discomfort, pain, or orthopedic deformities.[1] It is generally acknowledged that the physiological demand on the worker is increased during standing tasks. Cardiovascular demand is increased (manifested as increased heart rate and diastolic blood pressure), as is demand for continuous static contraction of the lower extremity and back muscles which maintain erect posture.[2] In the absence of leg movement, blood and other tissue fluids tend to accumulate in the legs, causing swelling and varicose veins (enlarged twisted veins most commonly observed in the lower extremity).[2] Fatigue and discomfort associated with standing tasks are attributable to insufficient return flow of the venous blood and static muscular effort.[3]

A few studies have examined the relationship between lower extremity musculoskeletal problems and prolonged standing postures.[4-5] A study conducted in Switzerland found that saleswomen who stand during the entire workshift had significantly more complaints of pain in the legs, feet, and back compared to saleswomen who walk around.[4] In another study, two-thirds of 315 saleswomen studied had signs or symptoms of varicosis.[4] In addition, 28%4 of the saleswomen had pains in the legs, 50% had problems with their feet, 18% had low back pain, and 15% had knee problems.

In a United Kingdom study, the most common site of regularly occurring pain or discomfort among female department store and supermarket staff was the feet.[5] In department store staff, pain or discomfort in the hips, legs, and knees was predominantly considered to be associated with either prolonged standing or general work fatigue, as was the case when discomfort in the feet was recorded. In supermarket staff, pain in the hips, legs, and knees was attributed to prolonged sitting or standing and cramped workstations. There were significant differences between department store and supermarket staff who reported foot discomfort and those who did not. Those who reported foot discomfort reported spending a greater time standing, walking, and kneeling, and

Health Hazard Evaluation Report No. 90-251

less time sitting, than did those without foot discomfort. In addition, this study of women in the retail trade observed a dose-response relationship between "time on feet" during the working day and prevalence of regular pain or discomfort in the feet. The prevalence of pain or discomfort was 48.2% among those spending more than 30% of their working day on their feet, while for those spending less than 30% the prevalence was 7%. Similar findings were found by Grandjean, who concluded that prolonged standing in one place is a common cause of ailments affecting the legs and feet of saleswomen.[3]

It is generally acknowledged that the harder and less "giving" the floor surface, the more stress is placed on the body. According to Konz, carpet is considered to be the best type of floor surface because it provides resilience; metal gratings are the worst since they not only have little resilience but also have minimum surface area, thus acting as knives.[2] A study by Redfern and Chaffin examining the effects of different floor surfaces, including concrete, seven types of mats, and a visco-elastic shoe insert found that the concrete floor and the hard mat consistently had the highest discomfort ratings.[1] The lowest ratings were shared by the 3/8" thick rubber mat, the tri-laminate mat, and the shoe insert. Although this study was limited by a small study population, it found significant levels of fatigue and discomfort in different areas of the body among workers who are required to stand for prolonged periods of time. The feet had the highest discomfort rating, followed by the ankle. How effective a type of floor is in relieving fatigue associated with prolonged standing appears to be a function of its hardness. However, a floor surface can be too soft. This is evidenced by the uneven mat, perceived to be the softest, yet receiving relatively high tiredness ratings.

VII. RESULTS AND DISCUSSION

Medical

Demographic and Work History Characteristics

Demographic and work history characteristics of the printers and news staff are presented in Table 1. On the average, printers were older than the news staff, by 19 years. The number of years worked at the WSJ by printers and news staff was similar. Proportionally, the number of females to males was greater among the news staff than among the printers. Self-reported heights for printers and news staff were similar.

Lower Extremity Discomfort

Five (25%) of the printers and 1 (13%) news staff person reported new onset lower extremity discomfort following the change in composing room floor surface. Perhaps because of the small number of people in the

Health Hazard Evaluation Report No. 90-251

study, this difference (relative risk [RR]=2.00, 95% confidence intervals [CI]: 0.27, 14.55) was not statistically significant. New onset LE problems included tiredness, burning and/or pain of the feet, cramps in the lower legs, and arthritis of the knee.

Twelve (60%) of the printers and 2 (25%) of the news staff reported lower extremity discomfort (either new onset or aggravation of a pre-existing condition) following the change of floor (RR=2.40, 95% CI: 0.69, 8.40). Again, this difference was not statistically significant.

Finally, 11 (55%) printers and 1 (13%) news staff person reported current lower extremity discomfort/pain (either new onset or an aggravation of a pre-existing condition) (RR=4.40, 95% CI: 0.67, 28.71). LE problems include tiredness and pain in the feet and lower legs, pain and swelling in the knee, and hip pain. Two employees were diagnosed with arthritis of the knee soon after the change in floor. One employee reported a history of hammertoes and ingrown toenails, one reported bunions on both feet, and another reported a history of a tumor removed from the foot. All of the printers and the news staff person with current LE discomfort/pain reported that their level of discomfort/pain decreased with introduction of the cushioned mats. Printers reported that the sit-down breaks during downtime also reduced LE discomfort/pain during the workshift. Many also reported that wearing shoes with cushioned soles reduced their level of discomfort.

Ergonomic

The current floor surface is a Roppe square-design rubber tile with foot cushions positioned in front of workstations associated with static standing (i.e., the dump area, ready bank, and make-up banks). The foot cushions are marketed as "anti-fatigue matting" and are open-celled mats composed of 3/32" thick top tile on a 3/8" thick foam rubber base. These mats have beveled edges on all four sides (eliminating a potential trip hazard). Several employees reported that the mats have a tendency to move over the day. Reportedly, maintenance workers are supposed to re-position the mats daily, but it was not being done consistently. A few areas where static standing occurs lacked cushioned mats. These areas include the dump area, the camera area, the opaque room, and the area where paper positives are made. The Roppe rubber tile design is reported to have a static coefficient of friction of 0.73, which indicates slip resistance. A slip resistance of not less than 0.50 has been traditionally recognized as not hazardous. The Roppe rubber tile has a tensile strength of 950 pounds per square inch, which indicates rupture resistance, elasticity, and resilience. This flooring surface replaced an industrial-grade woven-loop carpet with a 3/8" foam pad.

The 11 make-up banks are made of wood and are not adjustable. Each 2-sided table accommodates 6 pages of the WSJ, 3 on each side. The work surface has an approximate 30 degree slant. The distance from the

Health Hazard Evaluation Report No. 90-251

ledge, where the plastic page carrier rests, to the floor ranges from approximately 36 to 37 inches. The height from the top of a bank to the floor is approximately 50 to 51 inches. Extending across the top of each bank is a piece of glass where papers are clipped; this adds approximately 10 inches to the height of the workstation.

Although the make-up banks are not adjustable, it was reported that it is possible to raise the page by the use of blocks placed in the trough of the make-up bank. Two blocks are required to raise each page approximately 4 inches diagonally and 2 inches vertically. Reportedly, one of the tallest employees occasionally uses them. One of the shortest employees prefers a particular make-up bank because it is slightly shorter than the others.

A review of the videotapes and photographs revealed potential ergonomic hazards. Printers work at a standing work station. The tasks performed by printers have both dynamic (i.e., walking) and static (i.e., standing in one place) components. The major task involving static standing postures is making up the pages at the banks, especially when the pages are being made up initially for the 2-star edition. Generally, static standing postures were maintained for approximately 5-10 minutes and were interrupted by walks to other areas of the composing room, such as the ready bank and camera. In addition, short-statured printers were observed standing on their tiptoes in order to work on the top of the page.

In addition, employees were noted to be leaning on the front edge and sides of the make-up banks, resting on their elbows and other body parts. The front edge and sides of the make-up banks can act as a contact hazard because of their small surface area.

VIII. CONCLUSIONS

This investigation was limited by the small number of persons evaluated. Although differences in prevalence of lower extremity musculoskeletal problems between the printers and news staff were not statistically significant, more lower extremity problems were consistently reported by the printers than the news staff.

Static standing postures were observed among printers working at the make-up banks. Make-up banks did not accommodate the short-statured individual, as evidenced by employees standing on tiptoes. Static postures, when frequent or sustained, can be stressful and fatiguing to the musculoskeletal system and thus should be avoided.

IX. RECOMMENDATIONS

In operations where a standing workplace is used for a majority of the shift, it is desirable to avoid static postures requiring continuous muscle contraction. The following recommendations are offered to minimize static standing requirements:

Health Hazard Evaluation Report No. 90-251

1. Provide the option for sitting or standing at make-up banks. Work on large-size products or drawings lends itself well to a sit/stand operation; provision for sitting considerably reduces static loading on the legs and backs of workers as they lean over large drawings or negatives.[6]

2. Install foot rests 4-6 inches above the floor to allow workers to rest one leg while standing. Workers should alternate legs often. Foot rests are believed to alleviate back stress as well as minimize foot fatigue. Foot rests are needed even if sit/stand provisions are made.

3. The front edge and the sides of the make-up banks should be padded to allow workers to lean against the bank while working. (It may be necessary to bolt the banks to the floor for stability.) Padded edges also eliminate the potential contact hazard associated with leaning on a small surface area.

4. Cushioned mats should be installed at the dump area, the camera area, the opaque room, and the area where paper positives are made, as well as in other areas where static standing occurs. Mats should be re-positioned prior to each workshift, or some type of semipermanent adhesive backing (e.g., velcro) should be used to keep the mats in place.

5. The dimensions of the make-up bank conform to those recommended by Woodson.[7] However, make-up banks adjustable within a range of 10-15% (4-6 inches in height) could better accommodate short-statured workers. It may be possible to provide sufficient adjustability by providing pull-out, lock-in-place work platforms, 4-6 inches high, for short-statured workers to stand on. Platforms should be covered with a mat to provide cushioning and improve comfort during use.

6. Shoes with well-cushioned insteps and insoles should be provided to improve comfort.

7. Dow Jones should continue to permit employees to sit during slack time.

Health Hazard Evaluation Report No. 90-251

X. REFERENCES

1. Redfern MS, Chaffin DB [1988]. The effects of floor types on standing tolerance in industry. In: Aghazadeh F, ed. Trends in ergonomics/human factors V. New York: Elsevier Science Publishers B.V.

2. Konz S [1983]. Work design: industrial ergonomics. 2nd ed. New York: John Wiley & Sons.

3. Grandjean E [1982]. Fitting the task to the man: an ergonomic approach. London: Taylor & Francis Ltd.

4. Couture L [1986]. Health problems related to standing. Quebec, Canada: Commission de la Sante et de la Securite du Travail du Quebec. CREF No. 86082901 [French].

5. Buckle PW, Stubbs DA, Baty D [1986]. Musculo-skeletal disorders (and discomfort) and associated work factors. Chapter 2. In: Corlett N, Wilson J, Manenica I, eds. The ergonomics of working postures. Philadelphia, PA: Taylor & Francis.

6. Eastman Kodak Co. [1983]. Ergonomic design for people at work. Vol. I. New York: Van Nostrand Reinhold.

7. Woodson WE [1981]. Human factors design handbook. New York: McGraw Hill.

XI. AUTHORSHIP AND ACKNOWLEDGEMENTS

Report prepared by: Deanna Letts, R.N., M.S.
Nurse Officer
Medical Section

Field ergonomic evaluation: David Nestor, M.S.P.T.
Research Physical Therapist

Ergonomic analyses: Katharyn A. Grant, Ph.D.
Industrial Engineer

Daniel J. Habes, M.S.E.
Industrial Engineer
Psychophysiology and
  Biomechanics Section
Applied Psychology and
  Ergonomics Branch
Division of Biomedical and
  Behavioral Science

Health Hazard Evaluation Report No. 90-251

Sherry Baron, M.D., M.S.
Medical Officer
Medical Section

Leslie Copeland, M.S.
Visiting Industrial Engineer
Medical Section

Originating office: Hazard Evaluations and Technical
Assistance Branch
Division of Surveillance, Hazard
Evaluations, and Field Studies

Report typed by: Elaine Moore

XII. DISTRIBUTION AND AVAILABILITY OF REPORT

Copies of this report may be freely reproduced and are not copyrighted. Single copies of this report will be available for a period of 90 days from the date of this report from the NIOSH Publications Office, 4676 Columbia Parkway, Cincinnati, Ohio 45226. To expedite your request, include a self-address mailing label along with your written request. After this time, copies may be purchased from the National Technical Information Service (NTIS), 5285 Port Royal, Springfield, VA 22161. Information regarding the NTIS stock number may be obtained from the NIOSH Publications Office at the Cincinnati address. Copies of this report have been sent to the following:

1. Dow Jones & Company, Dallas Texas

2. Dallas Typographical Union No. 173, Dallas, Texas

3. Occupational Safety and Health Administration (OSHA), Dallas, Texas area office

4. OSHA, Region VI

For the purpose of informing affected employees, copies of this report should be posted by the employer in a prominent place accessible to the employees for a period of 30 calendar days.

Table 1

Dow Jones & Company
Dallas, Texas
HETA 90-251

Demographic and Work History Characteristics

Characteristics	20 Printers	8 News Staff
Gender:		
Females	11	6
Males	9	2
Age (mean, sd[1]):		
Females	50 (4)	31 (8)
Males	55 (4)	42 (13)
All	52 (5)	33 (10)
# Years worked at WSJ (mean, sd):		
Females	9 (2)	8 (7)
Males	18 (14)	17 (19)
All	13 (10)	10 (10)
Height in inches (mean, sd):		
Females	64 (2)	64 (2)
Males	69 (3)	69 (1)
All	66 (4)	65 (3)
Range	60 - 72	60 - 70

[1] sd = standard deviation

HETA 89-299-2230
JULY 1992
US WEST COMMUNICATIONS (USWC)
PHOENIX, ARIZONA
MINNEAPOLIS/StPAUL, MINNESOTA
DENVER, COLORADO

NIOSH INVESTIGATORS:
THOMAS HALES, MD
STEVEN SAUTER, PhD
MARTY PETERSEN, PhD
VERN PUTZ-ANDERSON, PhD
LAWRENCE FINE, MD, DrPH
TROY OCHS, MS
LARRY SCHLEIFER, EdD
BRUCE BERNARD, MD, MPH

I. SUMMARY

In July 1989, the Hazard Evaluations and Technical Assistance Branch of the National Institute for Occupational Safety and Health (NIOSH) received a joint request from the Communication Workers of America (CWA) and US West Communications (USWC) to evaluate how the use of video display terminals (VDTs) affects the health of Directory Assistance Operators (DAOs). The primary concern of both USWC and CWA was the effect on the operators' musculoskeletal system. To address this primary concern, a cross-sectional study of 533 workers from five distinct job titles employed within three metropolitan areas (Phoenix, Minneapolis/St. Paul, and Denver) was conducted.

Assessment of the upper extremity (UE) musculoskeletal system utilized symptom questionnaires and physical examinations. Data on demographics, individual factors (medical conditions and recreational activities), work practices, work organization, and psychosocial aspects of work, including electronic performance monitoring, were obtained from all participants by questionnaire. For one of the employee groups, the total number of keystrokes per day was generated from company computer records. The physical workstation was assessed using checklists of workstation configuration, and postural data were obtained from employees while they operated their VDTs.

Two types of musculoskeletal outcomes were defined for analysis:

1) Potential Work-related UE Musculoskeletal Disorders (UE Disorders) defined by physical examination and questionnaire,
2) UE Musculoskeletal Symptoms (UE Symptoms) defined by questionnaire alone based on a cumulative score of symptom duration, frequency and intensity.

Associations between workplace factors and UE Disorders were assessed by multiple logistic models generated for each of the four UE areas (neck, shoulder, elbow, hand/wrists), and for any work-related UE musculoskeletal disorder. Associations between workplace factors and degree of UE Symptoms were assessed by multiple linear models generated for each of the four UE joint areas (neck, shoulder, elbow, and hand/wrist). The physical workstation and posture information was not included in these analyses due to methodological limitations described in Appendix C.

Five-hundred-thirty-three (93%) of 573 selected employees participated in the study. Fifteen employees were excluded from the analysis because they were employed at their current job less than 6 months. The mean age of the remaining participants was 38 years, and the mean seniority on the current job was 6.3 years. Seventy-eight percent of participants were female, and 74% described themselves as "white."

One-hundred-eleven (22%) met our case definition for potential work-related UE musculoskeletal disorders. Probable tendon-related disorders were the most common (15% of participants), followed by probable muscle-related disorders (8%), probable nerve entrapment syndromes (4%), ganglion cysts (3%), and joint-related disorders (3%). The hand/wrist was the area most affected (12% of participants), followed by the neck area (9%), the elbow area (7%), and the shoulder area (6%). Phoenix workers had the highest prevalence of disorders (25%), followed by Denver workers (21%), and Minneapolis/StPaul workers (17%). Loop Provisioning Center (LPC) employees had the highest prevalence of disorders (36%), followed by Recent Change Memory Administration Center (RCMAC) employees (25%), Directory Assistance Operators (DAO) (22%), Centralized Mail Remittance (CMR) employees (20%), and Service Representatives (SR) (6%).

The OSHA 200 Log data and concerns expressed by USWC and CWA suggested that the DAOs would have the highest prevalence of disorders. This expectation was not supported by our findings. Although 22% of DAOs had musculoskeletal disorders which were potentially work-related, this prevalence was not higher than two of the four comparison groups utilized in this study and was similar to the prevalence rates reported in some previous studies of VDT users.

The following variables had statistically significant associations in the final models ($p<0.05$) with at least one of the outcome measures, although most of these associations have small point estimates (odds ratios) or small portions of the total variance explained (R-squared).

Of the three demographic factors, female gender was associated with degree of neck and shoulder symptoms, and non-white race was associated with elbow disorders. Of the nine medical history factors, five were frequently reported and entered into the final models. A history of physician-diagnosed thyroid conditions was associated with hand/wrist disorders, and a history of physician-diagnosed rheumatoid arthritis was associated with degree of elbow and hand/wrist symptoms. Recreational activities were not associated with UE disorders or degree of UE symptoms.

For the nine work practice variables, use of bifocals at work was associated with neck disorders, while use of glasses or contact lenses at work was associated with degree of elbow symptoms. For the 10 work organizational factors, overtime in the past year was negatively associated with degree of shoulder symptoms, and increasing number of hours spent at the VDT workstation per day was negatively associated with degree of hand/wrist symptoms.

For the 29 psychosocial variables, seven were associated with UE Disorders: fear of being replaced by computers, jobs which required a variety of tasks, increasing work pressure, lack of a production standard, lack of job diversity with little decision making opportunity, high information processing demands, and surges in workload. Seven psychosocial variables were also associated with degree of UE Symptoms: four mentioned previously (increasing work pressure, lack of job diversity with little decision making opportunity, high information processing demands, and surges in workload), plus uncertainty about one's job future, lack of co-worker support, and lack of supervisor support.

None of the eleven electronic performance monitoring variables were associated with UE Disorders in the final models, but five variables were associated with degree of UE Symptoms. More UE musculoskeletal symptoms were experienced in individuals who perceived that the monitoring system: 1) caused less socializing with co-workers, 2) rarely helped work performance and motivation, 3) caused more supervisor complaints regarding work performance, 4) closely monitored their work quality, or 5) caused more work.

Information to estimate the total keystrokes per day was available for 174 (71%) directory assistance operators. Increasing total keystrokes per day was not associated with UE Disorders or UE Symptoms in the final models. The relative low number of keystrokes per day performed by DAOs (mean 15,950) limits the ability to generalize these results to CMR employees or other data entry employees who may perform up to 80,000 keystrokes per day.

Efforts to analyze the effects of employee postural and workstation factors were thwarted by methodological constraints (Appendix C).

Several sources of potential bias could have influenced the results and interpretation of this study, including study design limitations, disease misclassification, and exposure misclassification. The very high prevalence of disorders among LPC employees and the much lower prevalence of disorders among SR raises the possibility that many of the workplace factors associated with UE Disorders and UE Symptoms are surrogates for these job titles. The study's cross-sectional study design cannot determine whether self-reported working conditions were causally related to work-related musculoskeletal outcomes. For example, did the negative psychosocial variables cause a musculoskeletal disorder, or did the negative psychosocial variables result from acquiring a musculoskeletal disorder? In addition, a total of 72 independent variables were analyzed for associations with nine dependent variables opening the possibility for associations due to chance (Type I error). Further discussion of these and other biases are described in Section VI of this report.

> On the basis of this evaluation, NIOSH investigators concluded that a high prevalence of potential work-related musculoskeletal disorders and symptoms was observed in this study. Factors associated with these disorders included demographics, prior medical conditions, work practices, psychosocial aspects of the workplace, and electronic performance monitoring. A few of the associations are inconsistent to those reported in the literature. Almost all of the physical workstations observed in this study were of high ergonomic quality, therefore we could not evaluate its contribution to work-related upper extremity musculoskeletal disorders and upper extremity musculoskeletal symptoms. This study adds to the evidence that the psychosocial work environment is related to the occurrence of work-related upper extremity musculoskeletal disorders and upper extremity musculoskeletal symptoms. Recommendations to improve working conditions, and possibly prevent and control musculoskeletal disorders are contained in Section VIII of this report.

KEY WORDS: SIC 4813 (Telecommunications), video display terminals, office automation, ergonomics, psychosocial, work stress, electronic performance monitoring, keystrokes, musculoskeletal disorders, tendinitis, carpal tunnel syndrome.

II. INTRODUCTION

In July 1989, the NIOSH Health Hazard Evaluation program received a joint request from US West Communications (USWC) and the Communication Workers of America (CWA) to evaluate "...how work may impact the health of Directory Assistance Operators (DAO)." NIOSH limited the health outcome to the musculoskeletal system because this was the primary concern of USWC and CWA. In addition, to obtain comparison groups for many of the work organization and psychosocial aspects of work, the study group was expanded to include other employee groups utilizing video display terminals (VDTs) and performing keyboard tasks for at least six hours per day. The other employee groups included Service Representatives (SR), Loop Provisioning Center (LPC) employees, Recent Change Memory Administration Center (RCMAC) employees, and Centralized Mail Remittance (CMR) employees.

Site-visits of four potential participating cities were conducted and information was gathered regarding the type of workstations, job requirements, and health data. In May 1990, study protocols were distributed to USWC, CWA, and three individuals outside of NIOSH with expertise in the areas of VDT use and musculoskeletal disorders. The protocol was modified to incorporate many of the suggestions offered by the reviewers. During the six month period, June 1990 to December 1990, data were collected from the three metropolitan areas selected for participation (Phoenix, Minneapolis/St Paul, and Denver). At each location, upon completion of the physical examination, the NIOSH physician discussed the examination's findings with the individual employee. An interim letter was distributed to CWA and USWC in January 1991, and preliminary results were discussed with the joint CWA-USWC ergonomics committee in January, 1992.

III. BACKGROUND

During the past decade musculoskeletal problems attributed to VDT use have been reported in the United States and other countries.[1-7] Musculoskeletal problems among VDT users have been linked to workplace ergonomic demands (eg constrained postures) in numerous studies,[8-11] however other investigations have provided equivocal support for these findings.[12-16] In particular, the psychosocial work environment (eg job control, social support) of VDT users has received increasing attention, with many studies finding relationships of the psychosocial work environment with health complaints.[17-22] Few scientific studies, however, have examined the role of these two categories of risk factors interactively.[12-13]

The objectives of this study were to 1) determine the prevalence of potential work-related UE musculoskeletal disorders among five employee groups utilizing VDTs, 2) determine the association of demographic, individual factors, work practices, work organization, psychosocial factors, electronic performance monitoring, and keystrokes per day with these disorders, and 3) to suggest prevention strategies to control the occurrence and severity of these disorders.

IV. METHODS

A. CITY AND PARTICIPANT SELECTION

Disease prevalence among employee groups was estimated using the company-maintained injury and illness records (Occupational Safety and Health Administration [OSHA] 200 Logs). Based on these estimates, sample size calculations were performed to detect a disease prevalence difference between DAOs and the other employee groups of 2.5 using the standard alpha (0.05) and beta (power) (0.80).[23] The city selection was narrowed to locations employing at least 125 DAOs, with at least 25 individuals in each comparison group: SR, LPC, RcMAC, and CMR. Six metropolitan areas qualified. Three metropolitan areas were selected for study: two (Phoenix and Denver) with a relatively high CTD incidence, and one (Minneapolis/StPaul) having a relatively low CTD incidence (based on the OSHA 200 logs).

For three of the employee groups, RcMAC, LPC, and CMR, all employees working the day of the NIOSH site-visit were asked to participate. The remaining two employee groups, DAO and SR, had more employees working than the required sample size, so a random sample of employees working the day of the NIOSH visit was selected.

B. MUSCULOSKELETAL OUTCOMES

A self-administered questionnaire designed to elicit data on musculoskeletal symptoms of the upper extremity (UE) was distributed to participating employees. If discomfort had been experienced in the past year, more information was ascertained regarding the discomfort's onset, duration, frequency, and severity. All employees completing the questionnaire were offered a physical examination of their upper extremities. The examination consisted of inspection, palpation, passive movements, resisted movements, and a variety of maneuvers to define UE musculoskeletal conditions standardized through its use in other NIOSH evaluations (Table 1). Four physicians were trained to administer the UE examinations and were blinded to the individual's questionnaire responses.

Based on the questionnaire and physical examinations, two types of musculoskeletal outcomes were defined for analysis.

1. **Potential Work-related UE Musculoskeletal Disorders (UE Disorders)** defined by physical examination and questionnaire.

Study participants were divided into cases and non-cases of potential work-related UE musculoskeletal disorders according to the criteria listed in Table 2.

2. **UE Musculoskeletal Symptoms (UE Symptoms)** defined by questionnaire.

Given the fluctuations of musculoskeletal disorders over a 12-month period, a case definition which requires positive physical examination findings may cause false negative results. In addition, employees may over- or under-report the work-relatedness of their symptoms. Therefore, a second method of classifying the UE musculoskeletal disorders was generated using only the symptom questionnaire. A cumulative score of the discomfort's duration, frequency, and severity was calculated separately for the neck, shoulder, elbow, and hand/wrist (Table 3). The UE symptoms scores did not include work-related criterion. Because the response scales for the duration, frequency, and severity questions varied in terms of number of response options, they were standardized prior to summation.[34]

C. **DEMOGRAPHICS AND INDIVIDUAL FACTORS**

The age, race, and gender of all participants were ascertain by the questionnaire. In addition, the questionnaire asked the total number of hours per week spent on hobbies and recreational activities, and whether the participant had any of a number of physician-diagnosed conditions reported to be associated with carpal tunnel syndrome (rheumatoid arthritis, diabetes mellitus, thyroid disease, disk disease in the low back or neck, alcoholism, gout, lupus, and kidney failure).

D. **WORK PRACTICES AND WORK ORGANIZATION**

Work practice and work organization characteristics were assessed by questionnaire. The work practice variables included the use of glasses or contact lenses, bifocals, trifocals, and granny glasses; typing skill and technique; length of time sitting continuously in the chair; frequency of arising from the chair; and seniority on the current job. The work organization variables consisted of the number of overtime hours, co-worker use of the same workstation, task rotation, hours spent at the VDT workstation, hours spent typing, and the number and type of work breaks.

E.   PSYCHOSOCIAL

Several scales from a separate NIOSH questionnaire were incorporated into the present questionnaire to assess the psychosocial aspects of the work environment.[35] These scales have had extensive use in occupational stress research and prior NIOSH studies. In addition, items and scales from two surveys addressing the psychosocial work climate (the Job Characteristics Inventory[36] and the Job Diagnostic Survey[37]) were included in the questionnaire.

In total, psychosocial components of the questionnaire included 22 scales (such as job control, work pressure, workload, etc.), and four single-item variables (Appendix A). All scales were factor analyzed to assure uni-dimensionality and were further analyzed to determine their internal consistency using Chronbach alpha coefficients.[38] Additionally, three single item electronic performance monitoring questions answered by all participants were analyzed as part of this section.

F.   ELECTRONIC PERFORMANCE MONITORING

Employees who worked under electronic performance monitoring or productivity standards were asked to respond to a series of questions regarding its fairness, accuracy, and its effects on their work environment. These scales were adopted from a University of Wisconsin study of work monitoring among telecommunication workers.[39] In total, 24 electronic performance monitoring items were grouped into seven scales variables and four single-item variables (Appendix B). All scales were factor analyzed to assure uni-dimensionality and were further analyzed to determine their internal consistency using Chronbach alpha coefficients.[38] Two of the scales, composed of items which varied in terms of number of response options, were standardized prior to summation.[34]

G.   KEYSTROKE INFORMATION

The number of keystrokes per day could be estimated for the DAO employee group. As part of the electronic performance monitoring for DAOs, computers monitored the number of calls taken per day, the total "on-line" time, and the number of searches required to find the correct telephone number (search ratio). Given the search ratio, the number of first and subsequent searches per day could be calculated. The number of keystrokes performed during the first and subsequent searches were collected from the "SMART" computer monitoring program described below. Adding the number of keystrokes per day for the first and subsequent searches allowed an estimation of the number of keystrokes per day.

For the Denver DAOs, the SMART program was designed to provide feedback to operators regarding the efficiency of their number search strategy. Operators were notified upon initiation of SMART monitoring, which occurred for the complete day, on three separate days during the month. Among other things, the program calculated the precise number of keystrokes for the first search and all subsequent searches. Therefore, for the Denver DAO subgroup, precise information was available on the number of keystrokes per day.

H. PHYSICAL WORKSTATION AND POSTURAL MEASUREMENTS

The physical workstation was assessed using checklists of workstation configuration, and postural data were obtained from employees while they operated their VDTs. Efforts to analyze the effects of employee postural and workstation factors were thwarted by methodological constraints (Appendix C).

I. STATISTICAL ANALYSIS

The statistical analysis describe below is schematically represented in Figure 1.

Step 1: Independent variables were grouped into 6 sets: 1) demographic; 2) work practices, work organization, and individual factors; 3) psychosocial; 4) electronic performance monitoring; 5) DAO keystrokes per day; and 6) Denver DAO keystrokes per day. Univariate analysis of independent variables within each of the six sets was conducted to determine their association with UE Disorders and UE Symptoms, as described in section B above. Independent variables were then excluded from these sets and excluded from further statistical analysis if they did not appear to be associated with UE Disorders or UE Symptoms (p-value >0.1). P-values were calculated using Student's t-test, analysis of variance, likelihood ratio chi-square test, or Pearson's chi-square test, as appropriate.

Step 2: Within each of the six variable sets, all independent variables with a p-value ≤0.1 were entered into either logistic or linear models to examine their effects while controlling for the effects of other variables. Logistic modeling was used to examine effects on UE Disorders, while linear modeling was used to examine effects on UE Symptom scores. A backward elimination procedure was used in both types of models to remove non-associated variables (p-value >0.05).

Step 3: All independent variables surviving Step 2 were combined together for subsequent analysis. Again backward elimination removed non-associated variables (p-value >0.05). Because the sample size was smaller for the electronic performance monitoring, DAO keystrokes per day, and Denver DAO keystrokes per day variables (approximately 450, 174 and 37 participants, respectively), these variables were not included at this stage in order to maximize power and reliability.

Step 4a: Electronic performance monitoring variables remaining after Step 2 were combined with the variables from Step 3. Again backward elimination was used, first on the monitoring variables and then on all remaining variables, to remove non-associated variables (p-value >0.05).

Step 4b: Keystrokes per day for all DAOs remaining after Step 1, were combined with the variables from Step 3 and tested simultaneously. If the p-value for this test was less than 0.05, backward elimination removed non-associated variables.

Step 4c: Keystrokes per day for Denver DAOs, if significant after Step 1, were combined with the variables from Step 3, and tested simultaneously. If the p-value for this simultaneous test was less than 0.05, backward elimination removed non-associated variables (p-values greater than 0.05).

Associations are reported as odds ratios (OR) with a 95% confidence interval (95% CI) for the logistic models, and as partial R-squares (R^2) for the linear models. Odds ratios with a value of less than 1.05 are not reported. Many of the exposure (independent) variables, particularly the psychosocial and electronic performance monitoring variables, have scaled values. The OR for these scaled variables represent the increased risk of disease for one increment within the scale. For example, if the response options for an independent variable ranged from "1" to "5", the OR represent the risk of disease for those responding "2" compared to those responding "1", or "5" to "4", but not of "5" to "1".

V. RESULTS

Overall, 533 (93%) employees agreed to participate. The Denver employees, and the DAO employees had the highest rates of participation (97% for both, Table 4). Fifteen employees were excluded because they had been at their current job less than six months, leaving a total of 518 employees. The actual sample size of each final model was slightly smaller due to some employees not responding to all variables in the questionnaire.

Descriptive data for potential work-related UE musculoskeletal disorders and the six independent variable sets are reported below. Variables in the final models are reported together with the descriptive data for each of the six independent variable sets.

A. MUSCULOSKELETAL DISORDERS

Overall, 111 (22%) employees met our case definition for potential work-related UE musculoskeletal disorders (UE Disorders). Phoenix employees had the highest prevalence of disorders (25%), followed by Denver employees (21%), and Minneapolis/StPaul employees (17%) (Table 5). Loop provisioning center employees had the highest prevalence of disorders (36%), followed by RcMAC (25%), DAO (22%), CMR (20%), and SR (6%) (Table 5).

Probable tendon-related disorders (15%) were the most common type based on positive physical examination findings, followed by probable muscle-related [tension neck syndrome and neck trigger points (8%), probable entrapment neuropathies (4%), joint-related findings (3%), and ganglion cysts (3%)] (Table 6).

1. Neck

Overall, 9% of employees had neck disorders. Denver employees and the LPC employees had the highest prevalences, 11% and 17%, respectively (Table 7).

2. Shoulder

Overall, 6% of employees had shoulder disorders. Phoenix employees and the RcMAC employees had the highest prevalences, 8% and 9%, respectively (Table 8).

3. Elbow

Overall, 7% of employees had elbow disorders. Denver and Phoenix employees had similar prevalences, 9% and 8% respectively. The DAO and LPC employees had the highest prevalences, both 9% (Table 9).

4. Hand/Wrist

Overall, 12% of employees had hand/wrist disorders. All three metropolitan areas had similar prevalences (12% and 13%), while the LPC employees had the highest prevalences (20%) and the SR the lowest (1%) (Table 10).

B. DEMOGRAPHIC AND INDIVIDUAL FACTORS

The mean age of participants was 38 years, with a mean seniority at their job of 6.3 years. Most of the participants were female (78%), and 74% described their race as white. In the logistic models, non-white race was associated with elbow disorders [odds ratio (OR)=2.4, 95% confidence interval (95% CI)=1.2, 5.0] (Table 11). In the linear models female gender was associated with increasing neck and shoulders symptoms (R^2=0.03 for both) (Table 12).

Participating employees spent a mean of 12 hours per week on recreational activities or hobbies. This factor was not significantly associated with any of the five models for UE disorders (neck, shoulder, elbow, hand-wrist, and any UE area), or any of the four models for UE symptoms (neck, shoulder, elbow, and hand-wrist).

Thyroid conditions were reported by 26 (5%) of participants, while rheumatoid arthritis was reported by 32 (6%) (Table 13). Presence of thyroid disorders was a strong predictor of hand/wrist disorders (OR=3.9; 95% CI=1.5, 9.9) (Table 11). Rheumatoid arthritis was associated with increasing elbow and hand/wrist symptoms (R^2=0.01 for both) (Table 12).

C. WORK PRACTICES AND ORGANIZATION CHARACTERISTICS

Sixty-two percent of employees wore glasses or contact lenses at work, and 10% wore bifocals (Table 14). Forty-eight percent stated their typing skill as "medium" (between 30 and 60 words per minute), and 70% had a "touch type" technique (Table 14). The mean number of times per day arising from their chair was 12, while the median length of time sitting in the chair continuously was 1 to 2 hours (Table 14). Use of bifocals at work was associated with neck disorders (OR=3.8, 95% CI=1.5, 9.4) (Table 11). Use of glasses or contact lenses at work was associated with increasing elbow symptoms in the linear models (R^2=0.02) (Table 12).

Seventy-four percent of employees had worked overtime in the past year, 69% had co-workers utilizing their workstations, and 13% stated they rotated tasks during the workday (Table 15). A mean of 7.3 hours was spent at their VDT workstations, of which 7.0 hours was spent typing. There was a mean of 4.2 brief breaks and 2.5 longer breaks during the workday (Table 15). Overtime in the past year had a negative association with increasing shoulder symptoms (R^2=0.01), and increasing number of hours spent at the VDT workstation per day had a negative association with increasing hand/wrist symptoms (R^2=0.02). (Table 12).

D. **PSYCHOSOCIAL**

The mean and range scores for the 29 psychosocial variables are listed in Table 16. In the logistic models, seven variables accounted for 14 associations with UE Disorders: six for the neck, one for the shoulder, three for the elbow, one for the hand/wrist, and three for any upper extremity (Table 11). Fear of being replaced by computers was the most consistent, being associated with neck (OR=1.5, 95% CI=1.2, 2.0), shoulder (OR=1.5, 95% CI=1.1, 2.0), elbow (OR=1.5, 95% CI=1.1, 2.0) and upper extremity (OR=1.3, 95% CI=1.1, 1.5) disorders. Jobs requiring a variety of tasks were associated with neck (OR 1.4, 95% CI=1.1, 1.7) and upper extremity disorders (OR 1.2, 95% CI=1.0, 1.4). Increasing work pressure was associated with neck (OR=1.2, 95% CI=1.0, 1.3) and upper extremity disorders (OR=1.1, 95% CI=1.0, 1.2). Routine work lacking decision making opportunities was associated with neck (OR=1.6, 95% CI=1.3, 2.0) and elbow disorders (OR=1.4, 95% CI=1.1, 1.8), and high information processing demands was associated with neck (OR=1.3, 95% CI=1.0, 1.6) and hand/wrist disorders (OR=1.2, 95% CI=1.1, 1.4). Lack of a production standard was associated with neck disorders (OR=3.5, 95% CI=1.5, 8.3), while surges in workload was associated with elbow disorders (OR=1.2, 95% CI=1.0, 1.3).

In the linear models, seven variables accounted for 13 associations with degree of UE Symptoms: two for the neck, one for the shoulder, five for the elbow, and five for the hand/wrist (Table 12). Four of these seven variables also had associations in the logistic modeling analysis (work pressure, surges in workload, routine work, and high information processing demands). The three variables which had associations only with increasing UE Symptoms included: 1) uncertainty about one's job future [neck (R^2=0.01), elbow (R^2=0.02), and hand/wrist symptoms (R^2=0.02)], 2) lack of co-worker support [elbow symptoms (R^2=0.01)], and 3) lack of supervisor support [hand/wrist symptoms (R^2=0.01)].

E. **ELECTRONIC PERFORMANCE MONITORING**

Four hundred eighty-one (93%) participants reported computer monitoring for quantity of work performed, while only 42% reported computer monitoring for quality of work performed (Table 16). Four hundred fifty-seven (88%) participants reported the presence of a productivity standard (Table 16) and 57% of those reported it was fair (Table 17). None of the eleven monitoring variables were associated with UE Disorders in the logistic models. Five monitoring variables were associated with increasing UE Symptoms in the linear models

(Table 18). Perceptions that the monitoring system resulted in: 1) less socializing with co-workers ($R^2=0.01$), 2) rarely helping work performance and motivation ($R^2=0.02$), and 3) more supervisor complaints regarding work performance ($R^2=0.02$) were associated with increasing neck symptoms. Individuals who perceived that the computer closely monitored their work quality reported increasing shoulder and elbow symptoms ($R^2=0.03$ and 0.05, respectively), and individuals who perceived that the monitoring system resulted in more work reported increasing elbow symptoms ($R^2=0.06$). It is important to point out that several of the demographic, individual factors, work organization, and psychosocial variables drop out of the models when the electronic performance monitoring variables are included in the models (compare Tables 12 and 18).

F.   KEYSTROKES

Information to estimate the total keystrokes per day was available for 174 (71%) directory assistance operators (DAOs). Phoenix operators averaged the most keystrokes (16,832), followed by Minneapolis/St Paul (16,708) and Denver operators (14,534). Neither the logistic or linear modeling analysis for these 174 operators found an association between increasing total keystrokes per day and UE Disorders or increasing UE Symptoms in the final models.

The SMART program provided precise keystroke information for 37 Denver DAOs. These 37 employees averaged 13,943 keystrokes per day, and this variable was not a predictor for UE Disorders or increasing UE Symptoms.

VI.   DISCUSSION

Potential Work-Related Musculoskeletal Disorders

The overall prevalence of potential upper extremity work-related musculoskeletal disorders (by questionnaire AND physical examination) was 22%. Using similar case definitions, other NIOSH studies of high risk employees (meatpacking industry and supermarket scanning cashiers) found prevalence rates of 62% and 51%, respectively.[40,41]

NIOSH investigators have also conducted studies of musculoskeletal disorders among VDT operators. A previous NIOSH study defined potential work-related musculoskeletal disorders by questionnaire alone and found prevalence rates for the upper extremities to be 38%.[4] Prevalence rates based on questionnaires alone tend to be double the rates based on both questionnaires and physical examinations.[43,44] Other studies of newspaper employees utilizing

VDTs reported a prevalence of at least 26% for lower arm tendinitis or carpal tunnel syndrome, and 26% suffering from painful hands and wrists.[45,46] It appears, therefore, that the prevalence of UE symptoms and disorders at USWC is similar to the prevalence among VDT users studied by NIOSH researchers,[4,12] and others.[43,46] None the less, 22% of the USWC workforce met the NIOSH definition of a physician-diagnosed upper extremity musculoskeletal disorder. This study did not address the impact these disorders have on productivity and health care costs.

Phoenix employees had the highest prevalence of upper extremity disorders (25%), followed by those in Denver (21%) and Minneapolis/StPaul (17%). Prior analysis of the company maintained OSHA 200 logs found the same order; however, the rate difference between Phoenix and Minneapolis/StPaul was much greater in the OSHA 200 logs. In addition, the OSHA 200 logs suggested that DAOs were the only employee group utilizing VDTs having a problem with work-related UE musculoskeletal disorders. The present study, however, found a very high prevalence of upper extremity disorders among LPC employees, a relatively high prevalence among the RcMAC, CMR, and DAO employee groups, and a very low prevalence among SR. Why the OSHA 200 logs did not detect musculoskeletal problems in these other job titles was not addressed in this study.

The very high prevalence of disorders among LPC employees and the lower prevalence of disorders among SR raises the possibility that many of the workplace factors associated with UE Disorders or UE Symptoms are surrogates for these job titles. Although over 61 workplace variables (72 independent variables minus the 3 demographic, 6 individual factors, and 2 city and job title variables) were collected and analyzed for associations for hand/wrist disorders, current job title had the largest odds ratio. Other than for high information processing demands, our study was unable to identify workplace factors which account for the difference in hand/wrist disorders between employee groups. Other workplace factors could be accounting for these differences between employee groups but were not measured in this evaluation.

Several potential biases may have influenced the prevalence of potential work-related musculoskeletal disorders found in this study.

1) Employees who developed work-related musculoskeletal conditions could have left the workforce or transferred to other jobs resulting in an underestimation of disease prevalence; a "survivor bias" which can occur in any cross-sectional study.[47]

2) Part of the case definition for potential work-related musculoskeletal disorders relied on self-reports of symptoms occurring over the past year and whether they were "work-related." Given the common occurrence of musculoskeletal pain due to non-

occupational causes, work-relatedness may have been overestimated. On the other hand, the second part of our case definition required a positive physical finding on examination. Given the fluctuating nature of musculoskeletal disorders, a positive physical finding a number of months prior to our evaluation could have become negative at the time we did our evaluation. This would result in an underestimation of the prevalence. The net effect of these two potential causes of misclassification on the estimate of disease prevalence is unknown. The physical examination and the case definition utilized in this evaluation, while lacking validation, have been standardized through their use in other studies.[4,40,41,43]

3) This study utilized two separate measures of musculoskeletal outcome: potential work-related UE musculoskeletal disorders (UE Disorders) defined by physical examination and questionnaire, and a cumulative score of UE musculoskeletal symptom frequency, duration, and intensity (UE Symptoms) defined by questionnaire. Despite the former requiring the symptoms be "work-related" and have a positive physical examination, these two outcome measures were positively correlated [as the severity of the symptoms increased (increased frequency, longer duration, increased severity) the prevalence of UE disorders increased], and some of the predictors for increasing UE Symptoms were also predictors for UE Disorders (increasing work pressure for neck, routine work lacking decision making opportunity and surges in workload for elbow, and high information processing demand for the hand-wrist). Nevertheless, many employees scoring toward the higher ends of the symptoms scales did not meet our UE disorders criteria.

We believe the UE Disorders criteria is the stronger outcome measure because it provides a more specific, and more objective measure of self-reported symptom information. However, increasing UE Symptoms as an outcome measure has a few advantages over UE Disorders; a) inclusion of symptomatic employees whose positive physical findings may have resolved, b) avoidance of the relatively arbitrary definition of categorizing symptomatic employees into cases and non-cases, and c) allows for a continuous rather than dichotomous outcome.

As noted previously, UE Symptoms did not require the symptoms to be "work-related." The UE symptoms scores did not include a component of work-relatedness because employees who work in jobs where biomechanical risk factors are obvious may attribute musculoskeletal symptoms to their job more readily than those employed in jobs without obvious biomechanical risk factors. If this occurred, associations between predictor variables and symptoms could be exaggerated.

4) If the self-reported symptoms represent fatigue-related conditions (disease misclassification), one would expect the prevalence of symptoms to increase during the employees' workshift.

In the univariate analysis, there was no association between increasing neck, shoulder, elbow, or hand/wrist symptoms and the time into the workshift. In addition, there was no association between neck, elbow, hand/wrist or upper extremity disorders and the time into the workshift. There was, however, an association between shoulder disorders and study participation at the end of the workshift [compared to participation at the beginning and middle of the shifts (OR=1.5 and 4.4, respectively)]. Inclusion of this variable into the shoulder disorders model did not cause any of the other variable to change (Table 11). It is unclear why participants in the middle of their shifts had fewer shoulder disorders than participants at the beginning or end of their shifts.

5) Participation rates varied among employee groups (Table 3). LPC and RcMAC employees had the lowest participation. This lower participation rate could be due, in part, to the fact that their workload continued to accumulate during the time taken to participate in the study. In contrast, for the three other employee groups, particularly the DAOs, the time taken to participate in the study represented a "work-break." If employees in LPC and RcMAC without musculoskeletal symptoms were less likely to participate, this could over-estimate the disease prevalence in these employee groups. Given the generally high participation rates among all employee groups, however, the magnitude of this potential bias is probably small.

Other Limitations

A total of 72 independent variables [demographics (3), city and job title (2), individual factors (6), work habits (9), work organization (10), psychosocial (29), electronic performance monitoring (11), and keystrokes per day (2)] were analyzed in the univariate analysis (Step 1). Despite our criterion for inclusion or removal from a model being consistent with most scientific studies (p-value <0.05), the large number of independent variables opens the possibility for false positive associations due to chance (Type I error). For this reason, the term "statistical significance" has been generally avoided in this report. Distinguishing "causal" versus "chance" associations is aided by the 1) strength of the association, 2) consistency of the association with the reported literature, 3) the biological plausibility, and 4) specificity of the health outcome. We have considered these factors in the subsequent discussions of the associations found in this study.

In addition, the study's cross-sectional design can only identify associations; it cannot clearly distinguish cause vs effect. This point is especially important for the exposure variables which rely on self-reported perceptions of the work environment.

Demographics

In the linear models women were at higher risk for having neck and shoulder symptoms. Other studies have also found female gender as a risk factor for UE musculoskeletal disorders,[4,48-51] but most were unable to study men and women performing the same job. The current study found female gender as a risk factor in jobs where men and women performed the same job tasks. The logistic modeling analysis (UE Disorders), which required positive physical examination findings, failed to find this association, suggesting that women may consider their musculoskeletal symptoms more severe, report their symptoms earlier and more accurately, or have more non-occupational upper extremity usage than men.

Non-white race was associated with elbow disorders in the logistic model. In a univariate analysis of musculoskeletal disorders in the poultry industry, African-Americans were found to have a higher prevalence of upper extremity disorders.[52] However, when the ergonomic demands of the job were entered into a multiple logistic models, race was not a statistically significant factor in the poultry study.[53] As with many other associations in this study, the relationship between race and musculoskeletal disorders could represent a city-job title surrogate or statistical artifact.

Although advancing age has been reported to be a significant risk factor for musculoskeletal disorders in the general population,[50-51] this study, and others evaluating VDT workers, found no association between age and musculoskeletal disorders or symptoms in the final models. Survivor bias (described above) could be one possible explanation.

Individual Factors

Although recreational activities have been associated with musculoskeletal disorders,[52] this study, like most NIOSH studies, did not find recreational activities to be an important confounding variable for UE musculoskeletal disorders.[40] This potential confounding variable was controlled by collecting information on the total number of hours spent on recreational activities. In both the logistic and linear models, this factor was not significantly associated with work-related disorders or symptoms in the final models. For the logistic model, this finding may be due, in part, to our case definition, which excluded individuals previously injured in a symptomatic joint area or who had incurred the symptoms prior to employment at USWC. In addition, some employees may have benefitted from the conditioning effect resulting from certain recreational activities, canceling the detrimental effect some employees may have experienced.[54]

Participants responded to questions regarding physician diagnoses of medical conditions reported to be associated with carpal tunnel syndrome. Gout, kidney failure, lupus, and disc disease in the neck were reported in 2% or less of participants and were excluded from analysis. Rheumatoid arthritis, thyroid disease, disc disease in the low back, diabetes and alcoholism were reported more frequently (Table 13). Reporting a history of physician diagnosed rheumatoid arthritis was associated with increasing hand/wrist and elbow symptoms in the linear models, while a history of thyroid conditions was a predictor of hand/wrist disorders and was also a predictor for increasing neck symptoms. Given the low prevalence of participants meeting our case definition for carpal tunnel syndrome, this finding may suggest that thyroid conditions might be associated with soft tissue structures other than the median nerve. However, given the opportunity for Type I error in this study, other studies need to confirm this finding before conclusions can be drawn between thyroid conditions and hand/wrist disorders. Rheumatoid arthritis is known to affect the hand/wrist area, however effects on the elbow are less common and usually represent long-standing disease. Perhaps proximal radiation of pain from the wrist area could account for its association with elbow symptoms.

A potential bias could have influenced the associations found between medical conditions and musculoskeletal disorders and symptoms. Self reports of physician diagnosed medical conditions were accepted without confirming the condition in individual medical records. Neither the magnitude or direction of this potential misclassification bias is known.

Work Practices and Work Organization

The association between bifocal use and neck disorders found in this study has been previously reported.[55,56] Bifocal use while using a VDT causes more head movements from keyboard to screen, a backward declination of the head, and increased static loading of the neck muscles.[56] The use of bifocals has been reported to alter the elbow flexion.[56] It is possible that eyewear use was associated with awkward postural adjustments resulting in discomfort in the elbow region.

This study found no associations between UE Disorders or UE Symptoms and self-reported typing skill, typing technique, or hours per day spent at the VDT workstation in the final models. Selection criteria for employee groups required VDT use for at least six hours per day; consequently, the mean VDT use per day was 7.3 hours (SD 0.95) and 97% of individual participants utilized VDTs for at least 6 hours per day. Therefore, there was probably insufficient variation in the length of VDT hours per day to fully evaluate its association with UE Disorders or UE Symptoms. Other studies of VDT use have found dose-response relationships between hours of VDT use and neck and shoulder symptoms,[57] but this association has not been a consistent finding in the literature.

The finding that increasing typing hours per day were protective for hand/wrist symptoms could be due to asymptomatic employees volunteering for overtime, or conversely, symptomatic employees not volunteering for overtime or not being allowed overtime by their supervisors. Both these situations would result in more typing hours per day for the asymptomatic employees. Similarly, the finding that overtime in the past year was protective for shoulder symptoms is probably due to the same self-selection bias. This finding does not invalidate the association between increasing typing hours per day and hand/wrist symptoms identified in other studies.[4,42]

This study corroborates the findings of other studies regarding the lack of an association between UE disorders or symptoms and typing technique (hunt and peck vs touch typing).[4,42] Other studies have found associations between UE symptoms and not getting up from the workstation,[4] but this study found no associations in the final models with the length of time sitting in a chair continuously, or with the number of times arising from the chair per day. In addition, in this study other administrative controls aimed at preventing musculoskeletal disorders (rotating job tasks, providing more frequent work-breaks, and self-regulated work pace) were not protective. Despite the lack of an association between work-breaks and UE Disorders, there is considerable support for their effectiveness in other studies.[58-64]

The work practice and work organization variables were measured by questionnaire rather than direct observation. For some variables (eg. frequency of rest breaks, length of time sitting in the chair, number of times arising from the chair) recall bias could be introduced. The validity for some of these work practices and work organizational variables reported on the questionnaire is unknown, thereby reducing our ability to draw definitive conclusions about the importance, or lack of importance, of these variables.

Psychosocial

Fear of being replaced by computers was associated with four disorders (neck, shoulder, elbow, and any upper extremity), and uncertainty about the job future was associated with increasing symptoms in three areas (neck, elbow, hand/wrist). Clearly, for employees with UE Disorders and UE Symptoms, job security was an important issue. Unfortunately, the study's cross-sectional design cannot distinguish cause from effect. Are concerns about job security causing musculoskeletal disorders, or are concerns over job security due to having a musculoskeletal disorder? A longitudinal study could overcome these cause/effect study design limitations, but given the rapid technological advances in the telecommunication industry with the resulting changes in work environment, this type of study would be quite difficult. In either case, lack of job security has been related to adverse psychological effects and poor physical health in other studies.[65-67]

Work pressure was associated with musculoskeletal conditions in both models: neck and upper extremity disorders in the logistic model, and increasing neck, shoulder, elbow, and hand/wrist symptoms in the linear model. This finding supports earlier studies that found work pressure contributing to adverse health outcomes among VDT operators.[13,68-69] However, the modest strength of the association found in our models suggests that reducing the work pressure would have only a modest effect on the prevalence of upper extremity musculoskeletal disorders.

Jobs which require a variety of tasks were associated with neck and any upper extremity disorders while routine work lacking decision making opportunities was associated with increasing elbow and hand/wrist symptoms, and neck disorders. Table 16 indicates that, in general, work tended to be rated as quite routine. For this reason, we speculate that task variety may not have provided the relief from musculoskeletal conditions that would normally have been anticipated. In addition, some groups of VDT users with extremely varied tasks (eg newspaper reporters) have rates of disorders similar to the VDT users in this study.[4,42]

Information processing demands were associated with hand-wrist disorders and hand-wrist symptoms, and neck disorders in the logistic model. This association identifies a factor which has not been previously investigated as a cause of upper extremity musculoskeletal disorders or symptoms. Other studies examining the effects of information processing demands on the musculoskeletal system would be useful.

Lack of support from co-workers was associated with elbow symptoms, while lack of support from supervisors was associated with hand/wrist symptoms. Many researchers consider social support a powerful buffering mechanism to mitigate the effect of heavy work demands.[70-71] Non-occupational psychosocial stress factors were not ascertained in our evaluation. We cannot, therefore, determine if non-occupational psychosocial stress factors are confounding our findings.

The lack of a productivity standard was a risk factor for neck disorders. This finding supports the opinion that the presence of a production standard, alone, does not create a negative psychosocial environment. On the other hand, most of the employees without production standards worked in LPC. Therefore, the presence of a production standard is confounded by job title (lack of a production standard being a surrogate for job title).

Numerous studies have documented the association between psychological stress and health complaints. However, controversy exists as to whether these health complaints represent an actual increase in disease, an increase in reporting, or somatization. Although several studies have linked psychological stress and medical diseases (peptic ulcer disease,[72] coronary artery disease,[73,74] hypertension,[75,76] and infections[77,78]) only two of these studies addressed the role of psychological stress causing objective signs of upper extremity musculoskeletal disease.[18,20] Although the mechanism of effect has yet to be clearly delineated, this study points to the needs to address psychological factors, especially work pressure and job insecurity, in efforts to control musculoskeletal disorders among VDT workers.

Electronic Performance Monitoring

Electronic performance monitoring at USWC tracks an individual's performance, which is then used as a component of employee evaluations. Advocates justify its presence as a means to increase productivity, provide timely employee feedback, and generate objective data for employee evaluations.[79] Detractors argue, on the other hand, that computer monitoring may lead to stress by encouraging competition and unrealistic performance expectations, diminishing opportunities for social interaction, and invading privacy. In this study, the presence of computer monitoring alone was not associated with musculoskeletal disorders or symptoms. However, increased neck, shoulder, and elbow symptoms were reported by individuals who perceived the computer monitoring as: 1) closely monitoring their work quality, 2) making them work more, 3) rarely helping their work performance or motivation, 4) invading the social aspects of their job, or 5) resulting in negative feedback from their supervisor. Monitoring was most strongly associated with neck symptoms. If monitoring is causally associated with the development of symptoms, it may be possible, with further study, to administer monitoring systems without creating these adverse psychological states.

Keystrokes

In this study, the number of keystrokes per day was not a risk factor for UE musculoskeletal disorders or UE musculoskeletal symptoms in the final models. It must be remembered, however, that keystroke information was available only for DAOs; DAOs typed a mean of 15,950 keystrokes per day (range 11,304 to 22,875; std dev 2,410). This finding cannot be generalized to CMRs or other data entry employees who may perform up to 80,000 keystrokes per day.

Physical workstation

Most employee groups participating in the study had adjustable furniture to accommodate individual differences, and all renovated workstations were of high ergonomic quality. This lack of variance did not allow our study to evaluate the relationship between the physical workstation and UE musculoskeletal disorders or symptoms. Other studies have documented the importance of biomechanical (ergonomic) factors causing work-related musculoskeletal disorders in VDT workers.[8-11]

Statistical

In this study many of the independent (exposure) variables are moderately or highly related to each other. The variables in the final models represent those with the strongest associations with UE Disorders and UE Symptoms using our model selection techniques. However, many of the independent variables could be replaced by another highly related independent variable and account for almost the same amount of dependent variable variance.

Associations between workplace factors and UE musculoskeletal disorders and symptoms found in the final models were derived from the main effects of the independent variables. Relationships between independent variables may be investigated in future analyses. If these analyses alter our scientific interpretation of the data, both USWC and CWA will be notified.

VII. CONCLUSIONS

High prevalences of potential work-related musculoskeletal disorders and UE symptoms were observed in this study. Factors associated with these disorders or symptoms included some variables from all categories investigated (demographic factors, prior medical conditions, work practices, work organization, psychosocial aspects of the workplace, and electronic performance monitoring. Importantly, this study adds to the evidence that the psychosocial work environment can be associated with the occurrence of work-related UE musculoskeletal disorders and UE musculoskeletal symptoms. This association was maintained despite controlling for individual factors (demographics, prior medical conditions, work practices) and work organization characteristics. The limitations of this study must be noted: a) the failure of the findings to support our initial hypothesis that DAOs would have the highest prevalence of work-related musculoskeletal disorders, b) the difficulty determining causality in a cross-sectional study which utilizes self-reports of the work environment and health outcome, c) the large number of independent variables evaluated probably causing some false positive associations (Type I errors), and d) the complex interactions of the psychosocial variables. The association of musculoskeletal outcomes with multiple psychosocial factors was a principle finding of this study. While recommendations for work re-design can be offered based on these findings (See Section VIII - Recommendations) these recommendations are tempered by 1) the modest strength of many of the associations, 2) the methodological limitations described above, and 3) the lack of studies conducted to determine the effectiveness of these interventions.

VIII. RECOMMENDATIONS

A. Continue the joint USWC-CWA Ergonomics Committee. Involvement by top management and union officials demonstrates the commitment USWC and CWA has given to this subject and provides the motivating force for complete implementation of committee recommendations.

B. Continue the local employee-employer ergonomic committees. These committees can provide valuable insight into identifying new or existing hazards, suggest potential solutions, and provide feedback on the effectiveness of various interventions.

C. Continue with purchasing the workstation equipment selected for each employee group based on the recommendations of the USWC-CWA Ergonomics Committee. Our observation indicated that most of the employee groups had adjustable furniture to accommodate individual differences, and that most renovated workstations were of high ergonomic quality. However, a few employees lacked this equipment. Suggested ranges for this equipment can be found elsewhere.[80-82]

D. NIOSH recommends VDT workers have visual testing before beginning VDT work and periodically thereafter to ensure that they have adequately corrected vision to handle such work.[83] In addition, individuals who wear bifocals at work should be evaluated by an eye specialist to determine if the current lenses are appropriate for the job.

E. Despite not finding associations between many work organization variables and UE disorders or symptoms in the final models of this study, the literature suggests they remain important factors to prevent or reduce UE symptoms. Suggested measures to consider include: providing periods of time away from the VDT, allowing more frequent opportunities for employees to get out of their chair, encouraging employees to take more frequent short rest breaks, restructuring work to allow for some component of self-pacing, limiting unwanted overtime, and job rotation.

F. When making changes in the psychosocial work environment, one should consider the following factors:
1. Providing job security. Ambiguity could be reduced in matters of job security and opportunities for career development. Employees need to be clearly informed of promotional opportunities and mechanisms for improving skills or professional growth within the organization, as well as impending organizational developments that may potentially affect their employment.[84]
2. Studies addressing the causes of work pressure, and what interventions are successful at reducing work pressure.
3. Providing job diversity with decision making opportunities, while not overloading employees with an excessive variety of tasks.
4. Fostering co-worker and supervisor support.
5. Reducing information processing loads for employees with excessive demands.
6. Reducing surges in individual workload.

G. For employee groups where performance is electronically monitored, the monitoring should help employee work performance and facilitate positive supervisor and social relationships.

H. The mean number of keystrokes per day performed by the DAO did not seem excessive by comparative standards. For the employee groups where the number of keystrokes per day is much greater (CMR), consider alternative technologies (eg. optical scanners to read the check amount), or changing the work organization (eg. after visualizing the check amount, the operator could strike one key, notifying the computer that the check amount equals the billed amount). Whatever changes are made, their impact on the job's psychological strain needs to be considered.

I. Prompt evaluations of employees with musculoskeletal symptoms by a health care provider should be available without fear of employer reprisal. All recommendations for surgery should generally be based on two independent physician recommendations. Review by the employee's primary physician and the corporate medical director may be helpful. Guidelines for health care providers to evaluate and treat these disorders have been published.[85,86]

IX. REFERENCES

1. Hocking, B. Epidemiologist aspects of "repetition strain injury" in Telecom Australia. Med.J.Australia 1987;147:218-222.

2. Nakesaseko M, Tokunaga R, Hosokawa M. History of occupational cervicobrachial disorder in Japan. J.Human Ergology 1982;11:7-16.

3. Knave BG, Wibom RI, Voss M, Hedstrom LD, Bergqvist UVO. Work with video display terminals among office employees: 1. Subjective symptoms and discomfort. Scand.J.Work Environ.Health 1985;11:457-466.

4. NIOSH [1990]. Hazard evaluation and technical assistance report: Newsday, Inc., Melville, NY. Cincinnati: Department of Health and Human Services, Public Health Service, Centers for Disease Control, National Institute for Occupational Safety and Health, NIOSH Report No. HHE 89-250-2046, NTIS PB91-116251.

5. Heyer N, Checkoway H, Daniell W, Horstman S, Camp J: Self-reported musculoskeletal symptoms among office video display terminal operators. In: Sakurai H, Okazaki I, Omae K (eds): Occupational Epidemiology. Excerpta Medicine International Congress Series 889: Excerpta Medica, 1990; 225-258.

6. Rossignol AM, Morse EP, Summers VM, Pagnotto LD. Video display terminal use and reported health symptoms among Massachusetts clerical workers. J Occup Med 1987;29:112-118.

7. Hanahran LP, Moll MB: Injury surveillance. Ch. VIII. In: Baker EL (ed): Surveillance in Occupational Health and Safety. Am.J.Public Health 1989;79(suppl):38-45.

8. Duncan J, Ferguson D: Keyboard operating posture and symptoms in operating. Ergonomics 1974;17:651-662.

9. Hunting W, Laubli T, Grandjean E. Postural and visual loads at VDT workplaces: I. Constrained postures. Ergonomics 1981;24:917-931.

10. Maeda K, Hunting W, Grandjean E. Factor analysis of localized fatigue complaints of accounting-machine operators. J.of Human Ergology 1982;11:37-43.

11. Sauter SL, Schleifer LM. Work posture, workstation design, and musculoskeletal discomfort in a VDT data entry task. Human Factors 1991;33(2):151-167.

12. Pot F, Padmos P, Brouvers A. Determinants of the VDU operator's well-being. In Work With Display Units. B.Knave and PG Wideback (Eds) Amsterdam: North Holland 1986 (pp. 16-25).

13. Ryan AG, Hague B, Bampton M. Comparison of data process operators with and without upper limb symptoms. Community Health Studies 1987;20:63-68.

14. Sauter SL, Gottlieb MS, Jones KC, Dodson VN, Rohrer K. Job and health implications of VDT use: initial results of the Wisconsin-NIOSH study. Communications of the Assoc. of Computing Machinery. 1983;26:284-294.

15. Kemmlert K, Kilbom A, Milerad E, Wistedt C. Musculoskeletal trouble in the neck and shoulder: Relationships with clinical findings and workplace design effects. In: Proceedings on work with display units (pp. 174-177). Stockholm: Swedish National Board of Occupational Safety and Health, 1986.

16. Starr SJ, Shute SJ, Thompson CR. Relating posture to discomfort in VDT use. J.Occup.Med. 1985;27:269-271.

17. Frese M. Stress at work, coping-strategies and musculo-skeletal complaints. In Osterholz, U., Karmaus, W., Hullmann, B., Ritz, B. (Eds.) Work-related musculo-skeletal disorders, Proceedings of an International Symposium. Bonn: Projekttrager "Humanisierung des Arbeitslebens" 1987:121-39.

18. Sievers, K., Makela, M., Klaukka, T. Work stress and musculoskeletal disease (In Finnish with English summary). In Kalimo, E., Kallio, V., (Eds.) Juhlakirja Jaakko Pajula II. Helsinki: Kansanelakelaitos 1989:425-40.

19. Leino, P. Symptoms job stress predict musculoskeletal disorders. J. Epidemiol.Community Health 1989;43:293-300.

20. Linton SJ, Kamwendo K. Risk factors in the psychosocial work environment for neck and shoulder pain in secretaries. J.Occup.Med. 1989;31(no.7):609-613.

21. Dimberg L, Olafsson A, Stefansson E, Aagaard H, Oden A, Anderson GBJ, Hansson T, Hagert C. The correlation between work environment and the occurrence of cervicobrachial symptoms. J.Occup.Med. 1989;31(no.5):447-453.

22. World Health Organization. Work with visual display terminals: psychosocial aspects and health. J.Occup.Med 1989;31:957-968.

23. Fleiss, J. Statistical Methods for Rates and Proportions (2-nd ed.) Wiley & Sons, NY, NY. 1981.

24. Viikari-Juntura, D. Interexaminer reliability of observations in physical examinations of the neck. Phys.Ther. 1987;67:1526.

25. DeGowin EL, DeGowin RL [1983]. Bedside diagnostic examination. 4th ed. New York, New York: MacMillan Publishing Co., Inc., p.714

26. Lister, Graham [1984]. The Hand: diagnosis and indications. 2nd ed. New York, NY: Churchill Livingstone, p. 210.

27. Hoppenfeld, Stanley [1976]. Physical examination of the spine and extremities. Norwalk, CT: Appleton-Century-Crofts, p.127.

28. Lister, Graham [1984]. The Hand: diagnosis and indications. 2nd ed. New York, NY: Churchill Livingstone, p. 210.

29. Finkelstein, H. Stenosing tenovaginitis at the radial styloid process. J.of Bone and Joint Surg. 1930;12:509.

30. Mossman SS, Blau JN. Tinel's sign and the carpal tunnel syndrome. Brit.J.Ind.Med. 1987;294:680.

31. Phalen, GS. The carpal tunnel syndrome. Seventeen years' experience in diagnosis and treatment of 654 hands. J.of Bone and Joint Surg. 1966;48A:211-228.

32. Pecina MM, Krmpotic-Nemanic J, Markiewitz AD. Tunnel syndromes. Boca Raton, FL; CRC Press, 1991:169-173.

33. Labidus, P. Stenosing tenovaginitis. Surg.Clinics of N.America. 1953;33:1317-1347.

34. Dixon WJ, Massey Jr FJ. Introduction to statistical analysis. New York, NY: McGraw-Hill, Inc., 1969. page 31.

35. Hurrell JJ Jr, McLaney MA. Exposure to job stress- A new psychometric instrument. Scand. J. Work Environ. Health 1988; 14(Suppl1):27-28.

36. Sims PH Jr, Szilaqyi AD, Keller RT. The measurement of job characteristics. Academy of Management Journal 1976;19(2):195-212.

37. Hackman JR, Oldham GR [1980]. Work Redesign. Reading, MA: Addison-Wesley.

38. Chronbach LJ. Coefficient alpha and the internal structure of tests. Psychometrika 1951;16:97-334.

39. Smith MJ, Caragon P, Sanders KJ, Lim SY, LeGrande D. Employee stress in jobs with and without electronic performance monitoring. Applied Ergonomics 1992;23(1):17-28.

40. NIOSH. [1989] Health hazard evaluation 88-180-1958: John Morrell & Co., Sioux Falls, SD. Cincinnati, OH: U.S. Department of Health and Human Services, Public Health Service, Center for Disease Control, National Institute for Occupational Safety and Health, NIOSH Report No. HHE 88-180-1958, NTIS Report No. PB-80-128-992.

41. NIOSH. [1990] Health hazard evaluation 88-344-2092: Shoprite Supermarkets, NJ and NY. Cincinnati, OH: U.S. Department of Health and Human Services, Public Health Service, Center for Disease Control, National Institute for Occupational Safety and Health, NIOSH HHE 88-344-2092.

42. NIOSH. [1991] Health hazard evaluation Interim report: Los Angeles Times, Los Angeles, CA. Cincinnati, OH: U.S. Department of Health and Human Services, Public Health Service, Center for Disease Control, National Institute for Occupational Safety and Health.

43. Armstrong TJ, et al. Occupational risk factors: cumulative trauma disorders of the hand and wrist. Final Contract Report to NIOSH #200-82-2507, Cincinnati, OH. 1985

44. Park CY, Cho KH, Lee SH. Occupational cervicobrachial disorder among international telephone operators. In: Sakurai H, Okazaki I, Omae K, ed. Occupational Epidemiology. Proceedings of the Seventh International Symposium on Epidemiology in Occupational Health, Toyko, Japan. Elsevier Science Publishers, 1990:273-276.

45. Rempel D, Lopes J, Davila R, Davis B. Cumulative trauma among visual display terminal users at a newspaper company. Hazard Evaluation System and Information Service, Department of Health Services, Berkeley, CA. 1987

46. Buckle, P. Musculoskeletal disorders of the upper limbs: A case study from the newspaper industry. In: Queinnec Y, Daneillou F, ed. Designing for Everyone. Proceedings of the Eleventh Congress of the International Ergonomics Association, Paris, France. Taylor and Francis Ltd, 1991:138-140.

47. Viikari-Juntura, E. Neck and upper limb disorders among slaughterhouse workers. Scand J Work Environ Health 1983;9:283-90.

48. Massey EW. Carpal tunnel in pregnancy. Obs and Gyn Survey 1978;33:145-148.

49. Sabour M, Fadel H. The carpal tunnel syndrome: A new complication ascribed to the pill. Am J Obs and Gyn. 1970;107:1265-1267.

50. Stevens JC, Sun MD, Beard CM, O'Fallon WM, Kurland LT. Carpal tunnel syndrome in Rochester, Minnesota, 1961 to 1980. Neurology 1988;38:134-138.

51. Lawrence RC, Hochberg MC, Kelsey JL, McDuffie FC, Medsger Jr TA, Felts WR, Shulman LE. Estimates of the prevalence of selected arthritic and musculoskeletal diseases in the United States. J Rheumatol 1989;16:427-441.

52. NIOSH. [1990] Health hazard evaluation 89-251-1997: Cargill Poultry Division, Buena Vista, GA. Cincinnati, OH: U.S. Department of Health and Human Services, Public Health Service, Center for Disease Control, National Institute for Occupational Safety and Health, NIOSH HHE 89-251-1997, NTIS Report No. PB-90-183-989.

53. Hales TR, Kiken S, Fine LJ, et al. Work-related Upper Extremity Musculoskeletal Disorders in Poultry Processing Plants. Paper presented at the Eleventh Congress of the International Ergonomics Association, Paris, France, July, 1991.

54. Simon, HB. Exercise, Health, and Sports Medicine. In: Rubenstein E, Federman DD, eds. Scientific American Medicine 1991, CTM, I, p 24.

55. Collins M, Brown B, Bowman K, Carkeet A. Workstation variables and visual discomfort associated with VDTs. Applied Ergonomics 1990;21:157-161.

56. Martin DK, Dain SJ. Postural modifications of VDU operators wearing bifocal spectacles. Applied Ergonomics 1988;19:293-300.

57. Walsh ML, Harvey SM, Facey RA, Mallette RR. Hazard assessment of video display units. Am Ind Hyg Assoc J. 1991;52:324-331.

58. Oxenburgh M. Musculoskeletal injuries occurring in word processing operators. Proceedings of the 21st Annual Conference of the Ergonomics Society of Australia and New Zealand. 1984:137-143.

59. Sauter SL, Schnorr TM. In: Rom WN, ed. Environmental and Occupational Medicine. Boston, MA. Little, Brown & Co., (in press).

60. Rogers, S. Recovery time needs for repetitive work. Seminars in Occupational Medicine 1990;2:19-24.

61. Sauter, S. Rest pauses and keystrok productivity and musculoskeletal discomfort. Presented at Human Factors Society Meeting, San Francisco, CA September 1991.

62. Zwahlen HT, Hartmann AL, Kothari N. How much do rest breaks help to alleviate VDT operator subjective occular and musculoskeletal discomfort? Proceedings of the International Scientific Conference: Work with Display Units 86, Part 1, Stockholm; 1986:503-506.

63. Sundelin G, Hagberg M, Hammarstrom U. The effects of pauses on muscular load and perceived discomfort when working at a VDT wordprocessor. Proceedings of the International Scientific Conference: Work with Display Units 86, Part 1, Stockholm; 1986:501-502.

64. Florine R, Cail S, Elias R. Psychophysiological changes in a VDT repetitive task. Ergonomics 1985;28:1455-1461.

65. Margolis BL, Kroes WH, Quinn RA. Job stress: an unlisted occupational hazard. J.Occup.Med. 1974;16:654-661.

66. Beehr TA, Newman JE. Job stress employee health and organizational effectiveness: a facet analysis, model, and literature review. Personnel Psychology 1978;31:665-699.

67. Cobb S, Kasl SV. Termination: the consequences of job loss. Washington, DC: DHHS Publication No. 77-224, 1977.

68. Smith MJ, Cohen BGF, Stammerjohn LW. An investigation of health complaints and job stress in video display operations. Human Factors 1981;23:387-400.

69. Caplan RD, Cobbs, French JRP. Job Demands and Worker Health. NIOSH Research Report, HEW Publication 75-160. Washington DC< US Department of Health, Education, and Welfare.

70. McKay CJ, Cooper CL. Occupational stress and health: some current issues. In: Cooper CL, Robertson, IT, eds. International Review of Industrial and Organizational Psychology. Chichexter: Wiley: 1987.

71. Cohen S, Wills T. Stress, social support and the buffering hypothesis. Psychological Bulletin 1985;98:310-357.

72. Feldman M, Walker P, Green JL, et al. Life events stress and psychosocial factors in men with peptic ulcer disease. Gastroenterology 1986;91:1370.

73. Ruberman W. Psychosocial Influences on mortality of patients with coronary heart disease. JAMA 1992;267:559-560.

74. Ruberman W, Weinblatt E, Goldberg JD, Chaudhary BS. Psychosocial influences on mortality after myocardial infarction. NEJM 1984 1984;311:552-559.

75. Schnall PL, Pieper C, Schwartz JE, et al. The relationship between job strain, workplace diastolic blood pressure, and left ventricular mass: Results of a case-control study. JAMA 1990;263:1929-1935.

76. Mathews KA, Cottington E, Talbott ED, Kuller LH, Siegel JM. Stressful work conditions and diastolic blood pressure among blue-collar factory workers. Am J Epidemiol. 1987;126:280-290.

77. Cohen S, Pyrrell DAJ, Smith AP. Psychological stress and susceptibility to the common cold. NEJM 1991;325:606-612.

78. Graham NM, Douglas RM, Ryan P. Stress and acute respiratory infection. Am J Epidemiol 1986;124:389-401.

79. Office of Technology Assessment, Congress of the United States. Automation of America's Offices (OTA CIT-287). Washington, DC: U.S. Government Printing Office, 1987.

80. American National Standards Institute. American National Standards for Human Factors Engineering of Visual Display Terminal Workstations. Santa Monica, CA: Human Factors Society, 1988.

81. Kroemer, KHE. VDT Workstation Design. In: Helander M, ed. Handbook of Human-Computer Interaction. Amsterdam: Elserier Science Publishers, B.V. (North Holland), 1988.

82. Sauter SL, Chapman LJ, Knutson, SJ. Improving VDT Work; Causes and Control of Health Concerns in VDT Use. Laurence, KS. The Report Store, 1986.

83. NIOSH [1984]. Congressional testimony: statement of J. Donald Millar, M.D., Director, National Institute for Occupational Safety and Health, Centers for Disease Control, Public Health Service, Department of Health and Human Services, before the Subcommittee of Health and Safety, Committee on Education and Labor, House of Representatives, May 15, 1984. NIOSH policy statements. Cincinnati, OH: U.S. Department of Health and Human Services, Public Health Service, Center for Disease Control, National Institute for Occupational Safety and Health.

84. Sauter SL, Murphy LR, Hurrell JJJr. Prevention of work-related psychological disorders. A national strategy proposed by the National Institute for Occupational Safety and Health (NIOSH). American Psychologist 1990;45:1146-1158.

85. Hales TR, Bertsche P. Management of upper extremity cumulative trauma disorders. AAOHN Journal 1992;40(3):118-128.

86. Rempel DM, Harrison RJ, Barnhart S. Work-related cumulative trauma disorders of the upper extremity. JAMA 1992;267(6):838-842.

X. AUTHORSHIP AND ACKNOWLEDGEMENTS

Report Prepared by: Thomas Hales, MD
Regional Medical Consultant
NIOSH Region VIII

Steven Sauter, PhD
Chief, Motivation and Stress Research Section
Applied Psychology and Ergonomics Branch

Field Assistance: Linda Cocchiarella, MD, MPH
Occupational Medicine Resident
Cook County Hospital, Chicago

Marile DiGiaccomo
Secretary
NIOSH Region VIII

Steve Douglass, MS
Graduate Student
Miami University

B.J. Haussler
Statistical Assistant
Statistical Services Section
Surveillance Branch

Urs Hinnen, MD, MPH
Visiting Research Scientist
Zurich, Switzerland

Troy Ochs, MS
Graduate Student
University of Cincinnati

Vern Putz-Anderson, PhD
Chief, Psychophysiology and Biomechanics Section
Applied Psychology and Ergonomics Branch

G. Robert Schutte
Health Survey Coordinator
Medical Section
Hazard Evaluation and Technical Assistance
  Branch

Joan Watkins, DO, MPH
Occupational Medicine Resident
University of Illinois, Chicago

Originating Office: Hazard Evaluation and Technical Assistance
                        Branch
                      Division of Surveillance, Hazard Evaluations
                        and Field Studies

The authors would like to thank Dr. Marvin Dainoff from Miami University, Dr. Barbara Silverstein from the University of Michigan, and Dr. Michael Smith from the University of Wisconsin for their help with the study's protocol.

In addition, the authors would like to thank the management of USWC and officers of the CWA for their cooperation and assistance during the data collection phase of the study. Finally, special recognition is due to Kathy Rockefeller from USWC and David LeGrande from CWA for their help and patience during the study.

XI. DISTRIBUTION AND AVAILABILITY OF REPORT

Copies of this report may be freely reproduced and are not copyrighted. Single copies of this report will be available for a period of 90 days from the date of this report from the NIOSH Publications Office, 4676 Columbia Parkway, Cincinnati, Ohio 45226. To expedite your request, include a self-addressed mailing label along with your written request. After this time, copies may be purchased from the National Technical Information Service (NTIS), 5825 Port Royal Road, Springfield, Virginia 22161. Information regarding the NTIS stock number may be obtained from the NIOSH Publications Office at the Cincinnati address.

Copies of this report have been sent to:

1. US West Communications, Denver, Colorado
2. Communication Workers of America, Washington, DC
3. OSHA Region VIII

For the purpose of informing affected employees, copies of this report shall be posted by the employer in a prominent place accessible to the employees for a period of 30 calendar days.

TABLE 1
Physical Examination Criteria for Various Medical Conditions
HETA 89-299, US West Communications

After performing each passive, active, and resisted maneuver the employee was asked to quantify the discomfort based on a five-point scale: 1=no pain, 2=mild pain, 3=moderate pain, 4=severe pain, and 5=the worst pain ever experienced. Maneuvers were considered significant if the discomfort score was ≥ 3.

NECK

Tension Neck Syn.: - Resisted flexion, or extension, or rotation. Trapezius palpation (spasm or trigger points).

Cervical Root Syn.: - Positive Spurling's maneuver.[24]

SHOULDER

Rotator Cuff Tendinitis:
- Active or resisted arm abduction $\geq$90 deg.
- Deltoid Palpation.

Bicipital Tendonitis: - Positive Yergason's maneuver.[25]

Thoracic Outlet Syn: - Positive hyperabduction and Adson's maneuvers[26-27]

ELBOW

Epicondylitis: - Medial or lateral epicondyle palpation.

Tendonitis: - Pain in the proximal 2/3 of the forearm on resisted wrist or finger flexion or extension.

Radial Tunnel Syn: - Positive Mill's maneuver.[28]

HAND-WRIST

Tendonitis: - Pain in the distal 2/3 of the forearm or hand on resisted wrist or finger flexion or extension.

deQuervain's Dis.: - Positive Finkelstein's maneuver.[29]

Carpal Tunnel Syn.: - Positive Tinel's and Phalen's maneuvers.[30-31]

Guyon Tunnel Syn.: - Positive Guyon Tinel's maneuver.[32]

Ganglion cysts: - Presence of ganglion cysts.

Joint-related: - Decreased MCP, or PIP range of motion ($\leq$100 deg.)

Trigger Finger: - Locking of finger in flexion or palpable tendon sheath ganglion.[33]

TABLE 2
Criteria for Potential Work-Related Cases of Musculoskeletal Disorders
HETA 89-299, US West Communications

- Symptoms of pain, aching, stiffness, burning, tingling, or numbness,
- Symptoms occurred within the past year,
- No previous accident or trauma to the symptomatic joint area,
- Symptoms began after employment with USWC,
- Symptoms occurred on the current job,
- Symptoms lasted for more than 1 week, or occurred at least once a month,
- Posititve physical finding of the symptomatic joint area (Table 1).

Table 3
Cumulative Symptoms Score Calculation for the Neck
HETA 89-299, US West Communications

		Score
Duration: How long does this NECK problem usually last?		
	less than 1 hour	__1
	1 hour to 1 day	__2
	more than 1 day to 1 week	__3
	more than 1 week to 2 weeks	__4
	more than 2 weeks to 4 weeks	__5
	more than 1 month to 3 months	__6
	more than 3 months	__7
Frequency: How often have you had this NECK problem in the past year?		
	almost never (every 6 months)	__1
	rarely (every 2-3 months)	__2
	sometimes (once a month)	__3
	frequently (once a week)	__4
	almost always (daily)	__5
Intensity: On average, describe the <u>INTENSITY</u> of the NECK problem using the scale below.		
	No pain	__1
	Mild pain	__2
	Moderate pain	__3
	Severe pain	__4
	Worst pain ever in life	__5

Cumulative Score = Duration Score + Frequency Score + Intensity Score

NOTE: If no symptoms were experienced over the past year for a particular joint area, the response was scored a "0".

TABLE 4
Participation Rates
HETA 89-299, US West Communications

Job Title	PHX	MSP	DEN	TOTAL
DAO	112 (96%)	58 (98%)	75 (97%)	245 (97%)
SR	24 (100%)	24 (92%)	27 (93%)	75 (95%)
LPC	24 (71%)	22 (88%)	25 (90%)	71 (84%)
RcMAC	31 (94%)	18 (75%)	18 (95%)	67 (88%)
CMR	23 (92%)	24 (89%)	28 (100%)	75 (96%)
TOTAL	214 (92%)	146 (91%)	173 (97%)	533 (93%)

TABLE 5
Prevalence of Potential Work-Related Upper Extremity Musculoskeletal Disorders (UE Disorders) by City and Job Title
HETA 89-299, US West Communications

Job Title	PHX	MSP	DEN	TOTAL
DAO	27%	14%	20%	22%
SR	8%	0%	7%	6%
LPC	41%	18%	48%	36%
RcMAC	21%	29%	28%	25%
CMR	25%	30%	8%	20%
TOTAL	25%	17%	21%	22%

TABLE 6
Types of Musculoskeletal Conditions Identified on the Physical Examination
HETA 89-299, US West Communications

	#Cases	Total	Percentage
Probable Tendon-related	76	513	(15%)
Rotator Cuff Tendonitis	29	513	(6%)
Bicipital Tendonitis	2	516	(<1%)
Epicondylitis	25	515	(5%)
Proximal Tendonitis	27	516	(5%)
Distal Tendonitis	41	516	(8%)
deQuervain's Disease	10	516	(2%)
Trigger Finger	1	517	(<1%)
Probable Nerve Entrapment	21	513	(4%)
Cervical Root Syndrome	2	515	(<1%)
Thoracic Outlet Syndrome	2	515	(<1%)
Radial Tunnel Syndrome	4	516	(1%)
Carpal Tunnel Syndrome	4	517	(1%)
Guyon Tunnel Syndrome	14	517	(3%)
Ganglion Cysts	13	517	(3%)
Probable Joint Related	15	516	(3%)
Muscle Related	43	516	(8%)

TABLE 7
Prevalence of Potential Work-Related Neck
Musculoskeletal Disorders (Neck Disorders) by City and Job Title
HETA 89-299, US West Communications

Job Title	PHX	MSP	DEN	TOTAL
DAO	7%	5%	8%	7%
SR	8%	0%	4%	4%
LPC	14%	9%	28%	17%
RcMAC	10%	11%	22%	14%
CMR	0%	9%	4%	5%
TOTAL	8%	6%	11%	9%

TABLE 8
Prevalence of Potential Work-Related Shoulder
Musculoskeletal Disorders (Shoulder Disorders) by City and Job Title
HETA 89-299, US West Communications

Job Title	PHX	MSP	DEN	TOTAL
DAO	10%	0%	8%	7%
SR	0%	0%	0%	0%
LPC	5%	5%	8%	6%
RcMAC	7%	11%	11%	9%
CMR	12%	4%	0%	5%
TOTAL	8%	3%	6%	6%

TABLE 9
Prevalence of Potential Work-Related Elbow
Musculoskeletal Disorders (Elbow Disorders) by City and Job Title
HETA 89-299, US West Communications

Job Title	PHX	MSP	DEN	TOTAL
DAO	11%	4%	-12%	9%
SR	0%	0%	4%	1%
LPC	9%	0%	16%	9%
RcMAC	3%	6%	6%	5%
CMR	6%	0%	4%	3%
TOTAL	8%	2%	9%	7%

TABLE 10
Prevalence of Potential Work-Related Hand-Wrist
Musculoskeletal Disorders (Hand-Wrist Disorders) by City and Job Title
HETA 89-299, US West Communications

Job Title	PHX	MSP	DEN	TOTAL
DAO	16%	7%	13%	13%
SR	0%	0%	4%	1%
LPC	18%	14%	28%	20%
RcMAC	7%	24%	17%	14%
CMR	19%	22%	0%	12%
TOTAL	13%	12%	12%	12%

TABLE 11
Associations With Potential Work-Related Upper Extremity Musculoskeletal Disorders in the Logistic Regression Model
(Methods Section: Step 3)
HETA 89-299, US West Communications

	Odds Ratio	95% Confidence Interval[@]
NECK: (n=512)		
Use of bifocals	3.8	1.5 - 9.4
Lack of a productivity standard	3.5	1.5 - 8.3
Routine work lacking decision making opportunities	1.6	1.3 - 2.0
Fear of being replaced by computers	1.5	1.2 - 2.0
Job requires a variety tasks	1.4	1.1 - 1.7
High information processing demands	1.3	1.1 - 1.6
Increasing work pressure	1.2	1.0 - 1.3
SHOULDER (n=510)		
Fear of being replaced by computers	1.5	1.1 - 2.0
ELBOW (n=513)		
Race (non-white)	2.4	1.2 - 5.0
Fear of being replaced by computers	1.5	1.1 - 2.0
Routine work lacking decision making opportunities	1.4	1.1 - 1.8
Surges in workload	1.2	1.0 - 1.3
HAND-WRIST (n=511)		
Thyroid condition	3.9	1.5 - 9.9
Current Department[*] LPC	1.9	0.7 - 5.1
DAO	1.1	0.5 - 2.5
RcMAC	0.8	0.3 - 2.4
SR	<0.1	<0.0 - 0.6
High information processing demands	1.2	1.1 - 1.4
UPPER EXTREMITY (n=512)		
Fear of being replaced by computers	1.3	1.1 - 1.5
Job requires a variety of tasks	1.2	1.0 - 1.4
Increasing work pressure	1.1	1.0 - 1.2

[*] LPC = Loop Provisioning Center employees
DAO = Directory Assistance Operators
RcMAC = Recent Change Memory Assistance Center employees
SR = Service Representatives
All are compared to the CMR (Centralized Mail Remittance) employees

[@] The confidence intervals, like the hypothesis tests, should be viewed with caution because of the large number of independent variables evaluated.

TABLE 12
Associations With Potential Work-Related Upper Extremity Musculoskeletal Symptoms in the Linear Regression Model
(Methods Section: Step 3)
HETA 89-299, US West Communications

	R-Squared (R^2)
NECK (n=514)	
Gender (female)	0.03
Increasing work pressure	0.02
City	0.02
Uncertainty about job future	0.01
Thyroid condition	0.01
TOTAL MODEL	**0.11***
SHOULDER (n=514)	
Gender (female)	0.03
Increasing work pressure	0.02
Overtime in the past year	0.01 (neg)
TOTAL MODEL	**0.06**
ELBOW (n=513)	
Routine work lacking decision making opportunity	0.02
Uncertainty about job future	0.02
Surges in workload	0.02
Use of glasses or contact lenses	0.02
Increasing work pressure	0.01
Lack of co-worker support	0.01
Rheumatoid arthritis	0.01
TOTAL MODEL	**0.11**
HAND-WRIST (n=511)	
High information processing demands	0.02
Hours spend at the VDT station per day	0.02 (neg)
Uncertainty about job future	0.02
Routine work lacking decision making opportunity	0.01
Increasing work pressure	0.01
Lack of supervisor support	0.01
Rheumatoid arthritis	0.01
TOTAL MODEL	**0.11***

* Sum of partial R^2 typically does not equal total model R^2.
(neg) = Negative association

TABLE 13
Prevalence of Physician-Diagnosed Medical Conditions
HETA 89-299, US West Communications

Medical Conditions	Number	Prevalence
Rheumatoid Arthritis	32	6%
Thyroid Disorders	26	5%
Disk Disease in the Low Back	25	5%
Diabetes Mellitus	16	3%
Alcoholism	16	3%
Disk Disease in the Neck	9	2%
Gout	5	1%
Kidney Failure	6	1%
Lupus	0	0%

TABLE 14
Work Practice Characteristics
HETA 89-299, US West Communications

WORK PRACTICES — Prevalence

Work Practice	Prevalence
Wearing of glasses or contacts when using the VDT	62%
Wearing bi-focals when using the VDT	10%
Wearing tri-focals when using the VDT	3%
Wearing granny glasses when using the VDT	2%
Typing Skill: Slow (<30 words/minute)	39%
Medium (30-60 words/minute)	48%
Fast (>60 words/minute)	13%
Typing Technique: Hunt and Peck	23%
Touch	70%
Other	7%
Typical length of time sitting continuously in chair	
Less than 1/2 hour	7%
1/2 hour to 1 hour	24%
1 hour to 2 hours	52%
Greater than 2 hours	17%

	Mean	Range
Number of times per day arising from your chair	12	0-100
Seniority at current job (in years)	6.3	0.5-34

TABLE 15
Work Organization Characteristics
HETA 89-299, US West Communications

WORK ORGANIZATION		**Prevalence**
Overtime in the past year		74%
Overtime in the past month		48%
Overtime in the past week		29%
Do co-workers use your workstation		69%
Rotating tasks during the workday		13%
Ability to regulate work pace:	Never	29%
	Rarely	23%
	Sometimes	29%
	Often	8%
	Almost Always	11%

	Mean	Range
Hours per day spent at the VDT	7.3	0-10
Typing hours per day	7.0	0-10
Number of brief breaks per day	4.2	0-50
Number of longer breaks per day	2.5	0-7

TABLE 16
Psychosocial Variables*
HETA 89-299, US West Communications

	Mean	Range
Meaningful work	5.1	1-7
Control over amount and quality of work	3.1	1-5
Control over job related matters	2.5	1-5
Skill utilization	2.4	1-5
Control over work policy and materials	2.0	1-5
Participation in work decisions	2.0	1-5
Cooperation between union and mgmt on health issues	2.6	1-4
Job requires a variety of tasks	1.5	1-3
Job satisfaction	SV#	

	Mean	Range
Lack of friends and relatives (home) support	1.6	1-5
Little interaction with others	1.8	1-5
Sum of supervisor, co-worker and home support	2.1	1-5
Lack of supervisor support	2.3	1-5
Lack of co-worker support	2.3	1-5
Uncertainty about job future	2.3	1-5
Little interaction with co-workers	2.4	1-5
Customer hostility	2.5	1-5
Replaced by a computer	2.7	1-5
Boring work	2.8	1-5
Surges in workload	3.2	1-5
Workload	3.6	1-5
Increasing work pressure	3.8	1-5
High information processing demands	2.3	1-4
Sum of workload mental demands	2.6	1-4
Routine work lacking decision making opportunities	2.9	1-4
Work requires high mental demands	2.9	1-4

	Prevalence
Computer monitors quantity	93%
Computer monitors quality	42%
Presence of a productivity standard	88%

* Spaces between variables signify a change in the scale range or scale direction
\# Standardized value, see Methods section in text

TABLE 17
Electronic Performance Monitoring Variables*
HETA 89-299, US West Communications

	Prevalence	
Fair productivity standard	57%	

	Mean	Range
Control and accuracy of computer monitoring	2.4	1-4
Monitoring rarely helps work performance and motivation	1.1	1-4
Monitoring invades social aspects of the job	2.4	1-4
Monitoring results in more work	1.3	1-4
Monitoring results in negative feedback from supervisor	1.5	1-4
Most recent evaluation on work quantity	1.7	1-4
Most recent evaluation on work quality	1.6	1-4
Difficulty meeting productivity standard	1.6	1-3
Computer closely monitors work quantity	SV#	
Computer closely monitors work quality	SV	

* Spaces between variables signify a change in the scale range or scale direction
\# SV = Standardized value, see Methods section in text

TABLE 18
Monitoring Variable Associations With Potential Work-Related Upper Extremity Musculoskeletal Symptoms in the Linear Regression Model
(Methods Section: Step 4a)
HETA 89-299, US West Communications

NECK (n=393)   <u>R-Squared (R^2)</u>

Monitoring rarely helps work performance and motivation	0.02
Monitoring results in negative feedback from supervisor	0.02
Monitoring invades social aspects of the job	0.01
Gender	0.05
Increasing work pressure	0.02
City	0.02
TOTAL MODEL	**0.16***

SHOULDER (n=162)

Computer closely monitors work quality	0.03
Gender	0.04
TOTAL MODEL	**0.08***

ELBOW (n=149)

Monitoring results in more work	0.06
Computer closely monitors work quality	0.05
TOTAL MODEL	**0.11**

HAND-WRIST

    No electronic performance monitoring variables remained in model. Model variables same as in Table 11.

*   Sum of partial R^2 do not typically equal total model R^2

FIGURE 1
OUTLINE OF STATISTICAL ANALYSIS
HETA 89-299, U.S. WEST COMMUNICATIONS

	Outcome = Case/Non-Case (Potential Work-Related Musculoskeletal Disorders)	Outcome = Ordinal (Musculoskeletal Symptoms)
	Univariate Analysis $P<0.10$	Univariate Analysis $P<0$
Step 1	A. Demographic B. Work Practices/Work Organization Individual Factors C. Psychosocial Factors D. Electronic Performance Monitoring (EPM) E. DAO Keystrokes Per Day (KPD) F. Denver DAO KPD	A. Demographic B. Work Practices/Work Organization Individual Factors C. Psychosocial Factors D. Electronic Performance Monitoring (EPM) E. DAO Keystrokes Per Day (KPD) F. Denver DAO KPD
	Multiple Logistic Regression $P<0.05$	Multiple Linear Regression $P<0.$
Step 2	Separate Regression Models for A-F	Separate Regression Models for A-F
Step 3	One Regression Model for Variable A-C Surviving Step 2	One Regression Model for Variable A-C Surviving Step 2
Step 4a	One Regression Model for Variables A-C Surviving Step 3 Plus EPM (Variable D) Surviving Step 2	One Regression Model for Variables A-C Surviving Step 3 Plus EPM (Variable D) Surviving Step 2
Step 4b	One Regression Model for Variables A-C Surviving Step 3 Plus KPD (Variable E) Surviving Step 2	One Regression Model for Variables A-C Surviving Step 3 Plus KPD (Variable E) Surviving Step 2
Step 4c	One Regression Model for Variables A-C Surviving Step 3 Plus Den. KPD (Variable F) Surviving Step 2	One Regression Model for Variables A-C Surviving Step 3 Plus Den. KPD (Variable F) Surviving Step 2

APPENDIX A
PSYCHOSOCIAL VARIABLES
HETA 89-299, US West Communications

		Alpha Coeff
1.	Control over job related matters	0.67
2.	Control over amount and quality of work	0.68
3.	Control over work policy and materials	0.55
4.	Participation in work decisions	0.89
5.	Skill utilization	0.67
6.	Cooperation between union and mgmt on health issues	0.68
7.	Job requires a variety of tasks	0.62
8.	Meaningful work	0.64
9.	Increasing work pressure	0.80
10.	Increasing workload	0.76
11.	Surges in workload	0.82
12.	Replaced by a computer	SS#
13.	Little interaction with others	SS
14.	Little interaction with co-workers	SS
15.	Lack of supervisor support	0.86
16.	Lack of co-worker support	0.77
17.	Uncertainty about job future	0.73
18.	Lack of friends and relatives (home) support	0.80
19.	Sum of supervisor, co-worker, and home support	0.80
20.	Boring work	0.82
21.	Work requires high mental demands	0.83
22.	High information processing demands	0.58
23.	Work requires very little thinking	0.61
24.	Sum of workload mental demands	0.79
25.	Customer hostility	SS
26.	Presence of a productivity standard	SS
27.	Computer monitors quantity	SS
28.	Computer monitors quality	SS
29.	Job Satisfaction	SV@

* Alpha coeff. = Chronbach alpha coefficients[36]
\# SS = Single scale
@ SV = Standardized value, see Methods section in text

APPENDIX B
ELECTRONIC PERFORMANCE MONITORING VARIABLES
HETA 89-299, US West Communications

		Alpha Coeff*
1.	Fair productivity standard	SS#
2.	Control and accuracy of computer monitoring	0.59
3.	Monitoring rarely helps work performance and motivation	0.82
4.	Monitoring invades the social aspects of the job	0.77
5.	Monitoring results in more work	0.73
6.	Monitoring results in negative feedback from supervisor	0.72
7.	Most recent evaluation on work quantity	SS
8.	Most recent evaluation on work quality	SS
9.	Difficulty meeting productivity standard	SS
10.	Computer closely monitors work quantity	0.20
11.	Computer closely monitors work quality	0.36

* Alpha coeff. = Chronbach alpha coefficients[36]
# SS = Single scale

APPENDIX C
PHYSICAL WORKSTATION CHARACTERISTICS AND POSTURAL MEASUREMENTS

METHODS

Workstation checklists noted characteristics of chairs, screens, tables, keyboards, and documents on approximately 340 of the 533 (64%) participating employees (Table C-1). In addition, postural measurements (height, distance, and angle) of the neck and upper extremities were taken while employees were working at their workstations (Table C-2). Distances were measured with a ruler, and angles were measured with a goniometer in combination with a carpenter's level. Postural measurements were made for approximately one-half of the participating employees. Selection of employees to collect workstation checklists and angle measurements were not made at random, but rather for the convenience of the investigators. Therefore, some employee groups were over-sampled while others were under-sampled. No employees who participated in the questionnaire and physical examination portion of the study refused to have this information collected. Because of the limitations noted below, the reliability of the analyses linking these conditions to musculoskeletal disorders could not be assured, and thus modeling analyses are not reported.

RESULTS

1. Chair

   All of the employees participating in the study had wheels or casters on their chairs, and 99% had some adjustable seat-pan height mechanism. Over 50% of the chairs had back support tension and tilt adjustability and seat-pan tilt adjustability (Table C-3).

2. Keyboard

   All of the keyboards were detached from the VDT screen, and 82% had a separate keyboard table. Over 78% of the keyboards had the ability to adjust the height, tilt, front-back, and lateral position (Table C-4). The key configuration was 58% Qwerty, 40% Dvorak, and 3% numeric. A mouse option was present in 4% of participant's workstations. Only 35% had space available for wrist rests or support, and 20% of keyboard edges were sharp.

3. Screen

   Sixty-six percent of the VDT screens were located on a separate table, and 95% were positioned in the center of the operator's workstation. Over 50% of the screens had the ability to adjust the height, tilt, front-back, and lateral position (Table C-5).

4. Document

Forty-six percent of participants utilized a document while working at the VDT workstation. Sixty percent of these employees had the document located in the center of their workstation. Ninety-five percent of document holders had lateral adjustability, 69% had front-back adjustability, and 10% had height adjustability.

5. Table and Accessories

Twenty-two percent of tables had sharp edges, and 14% did not have an adequate knee envelope. Sixty percent of participating employees had a foot rest available.

Table C-6 lists the results from the postural measurements.

DISCUSSION

Several factors severely limit interpretation of the postural and physical workstation analysis. 1) Some of the associations may be the result of multi-colinearity (eg confounding) and/or other problems which could not be controlled for in the analysis; for example, electronic keyboard height adjustment mechanisms and inadequate knee envelopes were found only in the Minneapolis/StPaul Directory Assistance Operators. 2) Important sampling bias may have been introduced by measuring employees at only one point in time and by the fact that employees frequently re-adjusted and shared workstations; how reproducible the postural data at various times throughout the day is unknown. 3) Bias could be introduced by individuals with UE symptoms seeking new workstation equipment or changing their posture frequently to accommodate their discomfort. 4) For the employees studied, workstation equipment provided by USWC was, in general, ergonomically correct. Therefore, there may have not been sufficient variance for most physical workstations to adequately evaluate its relationship with UE musculoskeletal disorders.

TABLE C-1
PHYSICAL WORKSTATION CHARACTERISTICS
HETA 89-299, US West Communications

Chair:
- Arm Rests Present
- Back Support Height Adjustable
- Back Support Tension Adjustable
- Back Support Tilt Adjustable
- Seat Pan Tilt Adjustable
- Seat Pan-Back Tilt Linked
- Seat Pan-Back Tilt Link Mechanism
- Seat Pan Height Adjustability
- Seat Pan Height Mechanism (manual vs pneumatic)
- Swivel/Coasters/Wheels
- Pan Compression

Keyboard:
- Key Configuration (Qwerty or Dvorak or Numeric)
- Key Type [numeric vs mixed (alpha and numeric)]
- Detachable
- Separate Table
- Sharp Edges
- Wrist Support (Wrist Rests or Support)
- Lateral Adjustment
- Front-Back Adjustment
- Height Adjustment
- Height Adjustment Mechanism
- Tilt Adjustment
- Tilt Adjustment Mechanism
- Mouse

TABLE C-1 (cont.)
PHYSICAL WORKSTATION CHARACTERISTICS
HETA 89-299, US West Communications

Screen:
- Position (Side or Center)
- Separate Table
- Lateral Adjustment
- Front-Back Adjustment
- Height Adjustment
- Height Adjustment Mechanism
- Tilt Adjustment
- Tilt Adjustment Mechanism

Document:
- Present
- Position (Side or Center)
- Lateral Adjustment
- Front-Back Adjustment
- Height Adjustment

Table:
- Sharp Edges
- Knee Envelop Adequate

Accessories: Foot Rest

TABLE C-2
PHYSICAL WORKSTATION CHARACTERISTICS - POSTURAL VARIABLES
HETA 89-299, US West Communications

Height and Distance Measurements:

Screen: Floor to the center of the monitor's screen.
Document: Floor to the center of the document.
Table: Floor to the top of the table surface.
Keyboard: Floor to the home row keys.
Chair: Floor to the top of the seat pan.
Arm Rest: Floor to the highest point on the arm rest.
Eye: Floor to the fold at the corner of the eye.
Elbow: Floor to the olecranon while typing on the keyboard (Height A, Figure 1).
Popliteal height: Floor to popliteal area under thigh (Height J, Figure 1).
Pan Compression: Compression of the chair seat pan padding.

Eye to Screen Distance: Front of the cornea to center of the screen.

Elbow to Keyboard Discrepancy: Keyboard height minus the resting elbow height.

Angle Measurements:

Wrist Ulnar Deviation: One arm of the goniometer placed over the third metacarpal; the other arm was positioned over the midline of the forearm with the axis approximately over the capitate (Angle D, Figure C-1).

Wrist Extension: Goniometer arms adjusted to be contiguous with the dorsal surfaces of the hand (along the third metacarpal) and forearm (Angle C, Figure C-1).

Shoulder Flexion: Angle between the humerus and the vertical plane (Angle G, Figure C-1).

Elbow Flexion: Angle between the horizontal plane and the ulna (Angle H, Figure C-1).

Inner Elbow Angle: Angle between the bicep and forearm (Angle E, Figure C-1).

Eye Gaze: Gaze angle from the eye to the middle of the VDT screen with respect to the horizontal (Angle I, Figure C-1).

Head Tilt to VDT: Difference between the angle formed by the Frankfort plane and the horizon while looking straight ahead (eye landmarks were the tragon of the ear and the external canthus of the eye) (Angle B, Figure C-2), and the angle formed by the Frankfort plane and the horizon while looking at the display (Angle A, Figure C-2).

TABLE C-3
Physical Workstations Characteristics - Chair
HETA 89-299, US West Communications

STATIC VARIABLES   **Overall**
Chair:

Arm Rests Present	69%
Back Support Height Adjustable	31%
Back Support Tension Adjustable	68%
Back Support Tilt Adjustable	50%
Seat Pan Tilt Adjustable	68%
Seat Pan-Back Tilt Linked	95%
Seat Pan-Back Tilt Link Locking	44%
Seat Pan Height Adjustability	99%
Seat Pan Height Mechanism (manual vs pneumatic)	75% pneumatic
Swivel/Coasters/Wheels	100%

TABLE C-4
Physical Workstation Characteristics - Keyboard
HETA 89-299, US West Communications

STATIC VARIABLES   **Overall**
Keyboard:

Key Configuration (Qwerty or Dvorak or Numeric)	Q=58%, D=40%, N=3%*
Key Type [numeric vs mixed (alpha and numeric)]	96% mixed
Detachable	100%
Separate Table	82%
Sharp Edges	20%
Wrist Support (Wrist Rests or Support)	35%
Lateral Adjustment	97%
Front-Back Adjustment	83%
Height Adjustment	78%
Height Adjustment Mechanism (power assist)	16%
Tilt Adjustment	90%
Tilt Adjustment Mechanism (power assist)	14%
Mouse	4%

* Percentages do not add up to 100% due to rounding errors

TABLE C-5
Physical Workstations Characteristics - Screen
HETA 89-299, US West Communications

STATIC VARIABLES — Overall

Screen:
Position (Side or Center)	95% center
Separate Table	66%
Lateral Adjustment	60%
Front-Back Adjustment	72%
Height Adjustment	52%
Height Adjustment Mechanism (power assist)	24%
Tilt Adjustment	95%
Tilt Adjustment Mechanism (power assist)	0%

TABLE C-6
Physical Workstation Characteristics - Postural Variables
HETA 89-299, US West Communications

	Mean
Height and Distance Measurements:	
Screen height	40.4"
Document height	30.3"
Table height	28.0"
Keyboard height	28.9"
Chair height	18.8"
Arm rest height	26.7"
Eye height	44.6"
Elbow height	26.6"
Popliteal height	16.9"
Pan compression	0.8"
Eye to screen distance	26.8"
Elbow to keyboard discrepancy	<0.1"
Angle Measurements:	
Wrist ulnar deviation	174°
Wrist extension	152°
Shoulder flexion	29°
Elbow flexion	10°
Inner elbow angle	109°
Eye gaze	12°
Head tilt	6°

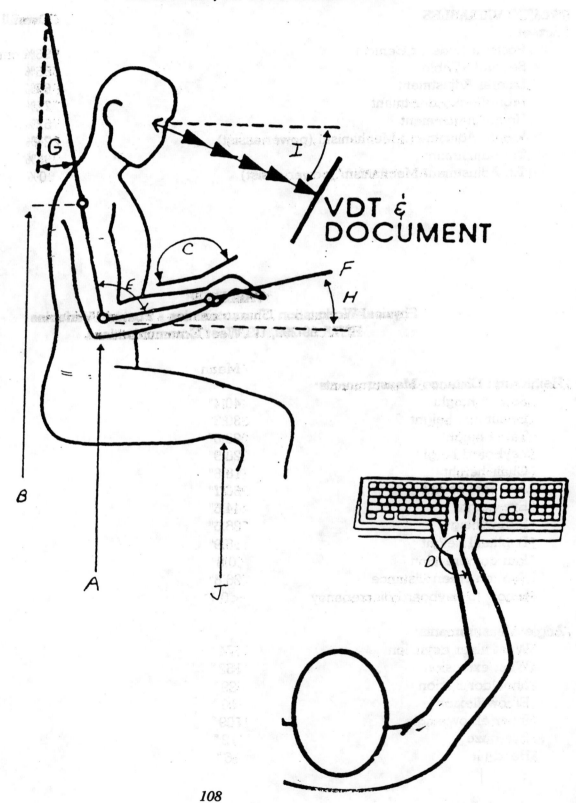

Figure C-1 (Appendix C)
Posture Measurements
HETA 89-299, US West Communications

Figure C-2 (Appendix C)
Posture Measurements
HETA 89-299, US West Communications

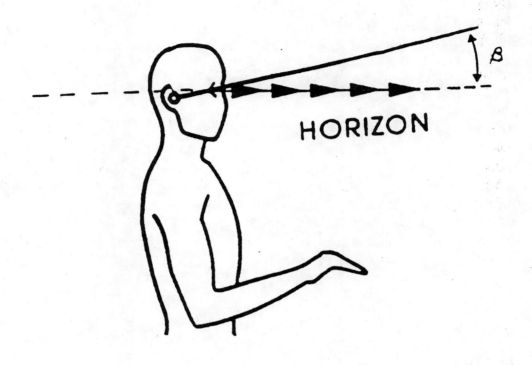

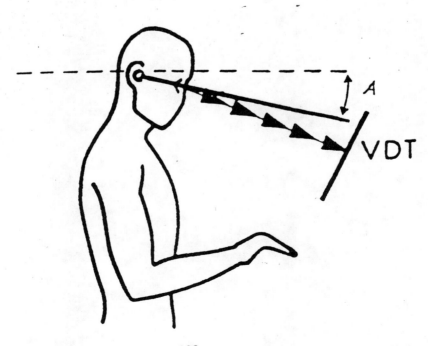

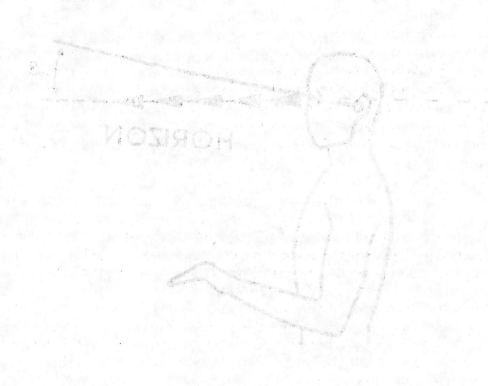

HETA 91-0208-2422
May 1994
Harley-Davidson Incorporated
Milwaukee, Wisconsin

NIOSH Investigators:
James McGlothlin, Ph.D, CPE
Sherry Baron, M.D., M.P.H.

I. SUMMARY

Researchers from the National Institute for Occupational Safety and Health (NIOSH) conducted initial[1] and follow-up health hazard evaluations of musculoskeletal disorders of the upper limbs and back at Harley-Davidson Incorporated, a motorcycle manufacturing company, over a 44 month period (January 1990 - August 1993). The objective of this evaluation was to identify job tasks in the flywheel milling department which may cause musculoskeletal injuries, and to provide recommendations to decrease and prevent such injuries.

NIOSH researchers reviewed the Bureau of Labor Statistics Log and Summary of Occupational Injuries and Illnesses (otherwise known as the OSHA 200 logs) and conducted an ergonomic evaluation of 4 jobs (2 flywheel milling, 1 truing flywheels, 1 flywheel balancing) in this department. Data gathered on the initial site visit in the flywheel milling area showed that repeated manual transport, placement, and removal of the flywheels between milling processes resulted in over 28,000 lbs. handled per 8-hour shift. In addition, repeated use of a hand-held power grinder to remove metal burrs from milled flywheels proved to be inefficient and potentially hazardous. Analysis of data from the flywheel truing job showed impact forces from the 5-lb. brass hammer repeatedly striking the flywheel ranged from 25,000 to 92,000 lbs. Analysis of the flywheel truing and balancing jobs showed potential risk for back injury, according to the revised NIOSH formula for manual lifting. Based on the initial evaluation, NIOSH provided recommendations to reduce these risk factors. The recommendations focused on reducing manual material handling, and increasing productivity and product quality. The company responded by forming an ergonomic committee consisting of management and labor to solve problem jobs in this department.

The committee focused on designing, redesigning, or eliminating jobs where musculoskeletal hazards were identified. Lighter flywheel castings from improved die-cast specifications, product flow, and better milling machines resulted in a reduction of flywheel handling to 17,500 lbs per 8-hour day. A customized 40-ton press eliminated the need for brass hammers, and an overhead lift eliminated manual handling of the assembled flywheel unit. During this five year period, 1989-1993, there was a reduction in the rate of cases of work-related musculoskeletal disorders involving lost or restricted workdays from 27.6 per 100 workers in 1989 to 12.5 per 100 workers in 1993. There was also a decrease in the severity rate of musculoskeletal

Health Hazard Evaluation Report No 91-0208-2422

disorders from 610 lost or restricted workdays per 100 workers in 1989 to 190 days in 1993. The process of evaluation and redesign of the flywheel department to reduce musculoskeletal disorders is presented as a model for reducing and preventing such disorders in this industry.

> On the basis of the information collected during this evaluation, NIOSH researchers determined that musculoskeletal hazards were significantly reduced following the development of an ergonomic program to improve the workstation design. However, potential for overexertion injuries to the back still exists in the flywheel assembly and truing jobs due to manual handling of flywheel assemblies. Recommendations for further reducing and preventing such hazards are presented in Section VI of this report.

Keywords: SIC 3751 (Motorcycle Manufacturing) Musculoskeletal Disorders, Manual Materials Handling, Cumulative Trauma Disorders, Metal Milling, Motorcycles, Ergonomics, Workstation Design, Engineering Controls.

Health Hazard Evaluation Report No 91-0208-2422

II. BACKGROUND

In the winter of 1990, NIOSH received a joint labor/management request from the Milwaukee facility of Harley-Davidson to evaluate musculoskeletal disorders of the upper limbs and back. Particular concern was expressed about the flywheel milling areas. The trigger for the NIOSH request was an increase in workers compensation costs due to an increased number of injuries. This increase occurred because of a number of factors including improved record keeping, increased awareness of work-related musculoskeletal injuries, an economic recovery resulting in the hiring of new workers, and other factors.

Based in part on the initial NIOSH report,[1] several ergonomic interventions were developed and implemented by the company. NIOSH then received a second joint labor/management request to evaluate these interventions. NIOSH representatives conducted a follow-up evaluation in May 1992, August 1992, October 1992, and August 1993. These visits generated information on the rates of reported musculoskeletal disorders and evaluated the ergonomic interventions that had already occurred and those that were ongoing.

A. Work-Related Musculoskeletal Disorders

Several case reports over the years have cited certain occupational and nonoccupational risk factors which give rise to musculoskeletal injuries.[2,3,4,5] However, only recently have epidemiologic studies (cross-sectional and case-control retrospective studies) been conducted that have examined the association between job risk factors (such as repetition, awkward postures, and force) and excess musculoskeletal morbidity.[6,7,8,9,10,11] These studies have identified relationships between these risk factors and the development of musculoskeletal disorders.

Upper Limbs

Work-related musculoskeletal disorders (WRMDs) of the upper limbs have been associated with job tasks that include: (1) repetitive movements of the upper limbs, (2) forceful grasping or pinching of tools or other objects by the hands, (3) awkward positions of the hand, wrist, forearm, elbow, upper arm, shoulder, neck, and head, (4) direct pressure over the skin and muscle tissue, and (5) use of vibrating hand-held tools. Because repetitive movements are required in many service and industrial occupations, occupational groups at risk for developing WRMDs of the upper limb continue to be identified.

Health Hazard Evaluation Report No 91-0208-2422

Evaluation of work-related risk factors which may cause upper limb WRMDs should be conducted in order to implement controls to reduce these risk factors. Engineering controls are the preferred method; however, administrative controls such as work enlargement, rotation, etc., can be used as an interim measure. Surveillance of WRMDs (including the use of health-care-provider reports) can aid in identifying high-risk workplaces, occupations, and industries and in directing appropriate preventive measures.[12]

Low Back Injuries

Occupational risk factors for low back injuries include manual handling tasks,[13] twisting,[14] bending,[17] falling,[15] reaching,[16] lifting excessive weights,[17,17,18] prolonged sitting,[18] vibration,[17,19] and job satisfaction.[20,21] Some nonoccupational risk factors for low back injury include obesity,[22] genetic factors,[23] and smoking.[24]

Control and prevention of job-related low back pain can be accomplished through the evaluation of jobs and the identification of job risk factors. Redesign of jobs can lead to the reduction of these risk factors and good job design initially will prevent back injuries. Multiple approaches such as job redesign, worker placement, and training may be the best methods for controlling back injuries and pain.[24]

B. **Workforce**

The NIOSH evaluation of this motorcycle manufacturing plant focused on the flywheel milling and assembly department, where milling, assembly and truing, and balancing of flywheels are done.

Pre-Intervention Evaluation

In January 1990, the plant employed about 500 workers and produced approximately 253 motorcycle engines and 170 motorcycle transmissions each day. There were 38 full-time workers in the flywheel milling department. Production was 24 hours per day, with two-to-three employees per shift working in the milling area and another five employees assembling, truing, and balancing the flywheels. Employees rotated through the truing task every 2-to-4 hours. Occasionally, these employees may have worked 10-to-12 hour days to keep pace with production demands.

Health Hazard Evaluation Report No 91-0208-2422

Post-Intervention Evaluation

In August 1993, the plant had increased employment and production by about 40% to about 700 hourly workers and produced approximately 340 motorcycle engines and 254 motorcycle transmissions a day. The flywheel milling department increased to 48 employees with two to three employees working in the milling area and seven assembling, truing, and balancing flywheels. Because of changes to the truing area (described below), employees did not need to be rotated.

C. Process Description

Milling of the flywheels consists of a series of steps: manually removing the flywheel from a supply cart; drilling and machine milling it; grinding off metal burrs; and inspecting, measuring, and placing the finished flywheel in a receiving cart. Each milling "cell" contains three to four milling machines, a drill press, and two-to-three worktables. Approximately two-three lbs. of metal are cut from each flywheel during the milling process.

After milling, the next phase is assembly of the flywheel unit. These components consisting of the gear and sprocket side of the flywheel, two connecting rods, bearings, and a crank pin, are assembled and sandwiched together by a "marriage press." After this, the unit is taken to the truing area for straightening and centering.

Truing of the flywheel was formerly done manually by mounting it on a fixture on top of a table, and manually rotating the flywheel to determine misalignment (a centering gauge is viewed by the operator to determine misalignment). In the initial NIOSH evaluation, when the misalignment area was found, the employee repeatedly struck the flywheel with a 5-pound brass-head hammer to straighten and center (true) the unit. In the follow-up evaluations, a 40-ton press performed the truing operation, and the hammers were eliminated. After the flywheel unit is trued, it is manually lifted and placed onto a cart, which is moved to the balancing area.

Balancing the flywheel unit is done by lifting it from a cart and placing it in a cradle in the balancing machine. The flywheel connecting rods are attached to balancing arms which rotate the unit at high speeds. A computer determines where holes are to be drilled to provide balance when the flywheel unit is operating at high speeds. Following this procedure, the flywheel is manually moved from the balancing machine to a cart. The weights of fully assembled post-milled flywheels (sprocket

Health Hazard Evaluation Report No 91-0208-2422

and gear) and their components (bearings, crank pin, and two connecting rods) are: large (FL) 32.5 lbs. (34 lbs. maximum weight), and small (XL) 25.6 lbs (26 lbs. maximum weight). The finished flywheel units are then moved from this department to the engine assembly department.

III. DESIGN AND METHODS

NIOSH researchers conducted an initial evaluation in January 1990 and four follow-up evaluations (May, August, and October 1992, and August 1993). The evaluation of the flywheel milling department included, a review of OSHA 200 logs, informal interviews with employees, and an ergonomic evaluation of jobs in the flywheel milling area.

A. Ergonomic Evaluation

Initial Evaluation

An in-depth ergonomic evaluation of the flywheel milling area was conducted during the initial survey consisting of: (1) discussions with flywheel milling employees regarding musculoskeletal hazards associated with their job, (2) videotaping the flywheel milling process, (3) biomechanical evaluation of musculoskeletal stress during manual handling of the flywheels, and (4) recording workstation dimensions. Two flywheel milling cells were evaluated.

Videotapes of the jobs were analyzed at regular speed to determine job cycle time, slow-motion to determine musculoskeletal hazards of the upper limbs during manual material handling tasks, and stop-action to sequence job steps and perform biomechanical evaluations of working postures. All video analysis procedures were used to document potential musculoskeletal hazards in performing the job.

Time and motion study techniques were used for the first phase of job analysis.[25] *Work methods analysis* was used to determine the work content of the job.[26] The second phase of job analysis was to review the job for recognized occupational risk factors for WRMDs. These WRMDs risk factors include repetition, force, posture, contact stress, low temperature, and vibration.[27,28] In addition, biomechanical evaluation of forces which are exerted on the upper limbs, back, and lower limbs of the worker while performing the task also was performed.[29] This two-phase approach for job analysis and quantification of forces which act upon the body during materials handling forms the basis for proposed engineering

Health Hazard Evaluation Report No 91-0208-2422

and administrative control procedures aimed at reducing the risk for musculoskeletal stress and injury.

After receipt of the initial NIOSH report in October 1990, the company conducted several meetings over a 1-2 year period to engineer out specific job hazards in the flywheel milling and assembly department. The meetings led to the systematic selection of equipment and process changes based on over 20 performance criteria. Some of these criteria were: reduction or elimination of the specific hazard (vibration from hand tools), user friendly controls, noise reduction, easy access for maintenance personnel, parts availability, cycle time, machine guarding, and machine durability.

Follow-up Evaluations

Four follow-up evaluations were conducted between May 1992 and August 1993. During these evaluations NIOSH researchers spoke with the safety director as well as the operators, managers and engineers involved in the redesign of the work processes in the flywheel department. An evaluation was also done on the changes made since the initial evaluation. Specific NIOSH activities during these follow-up evaluations included: (1) discussions with employees regarding changes in their job for musculoskeletal hazards, (2) videotaping the flywheel milling, truing, and balancing process, (3) reviewing company ergonomic committee activities on reducing job hazards in this department, (4) presenting education and training sessions on ergonomics to plant supervisors, engineers, and workers, and (5) reviewing records on reports of cases of musculoskeletal disorders.

Incidence rates of musculoskeletal disorders (the number of new cases per 100 workers) between 1987 and 1993 were determined using the OSHA 200 logs. All musculoskeletal problems including such conditions as sprains, strains, tendinitis, and carpal tunnel syndrome involving the upper extremities, neck, and back recorded on the OSHA 200 log were included in this analysis. Since it is often difficult to determine from the OSHA 200 logs whether a musculoskeletal sprain or strain is due to acute or chronic trauma, all of these events were included. Musculoskeletal contusions, which are likely to be more acute events, were not included. Data on the number of lost and restricted workdays for each case were also tabulated.

Information on the total number of hourly employees for each year was obtained and used to develop incidence rates. The incidence of musculoskeletal disorders was calculated for each year and each area of

Health Hazard Evaluation Report No 91-0208-2422

the body. Additionally, severity rates were calculated by examining the rate of lost or restricted workdays and the median number of lost or restricted workdays per case.

IV. RESULTS

A. Ergonomic Evaluation

Table 1 summarizes the initial ergonomic recommendations for the flywheel milling area made by NIOSH researchers in January 1990, and the actions completed by the company by the last follow-up evaluation in August 1993. This table shows that several actions were taken to address concerns about musculoskeletal injuries in the flywheel milling cell. Pre- and post- ergonomic intervention activities for the milling, assembly and truing, and balancing areas are summarized below.

Milling

Pre-Intervention Evaluation: Milling of the large (FL) flywheel (average weight 19.0), and the small (XL) flywheel (average weight 16.0), consists of 37 steps for the FL flywheel, and 25 steps for the XL flywheel. It was estimated that 28,175 lbs of flywheels were manually handled for the FL flywheel, and 18,980 lbs for the XL flywheel per 8-hour day. These total weights were derived by multiplying the average weight of the milled flywheel (17.5 lbs FL, and 14.5 lbs XL) times the average number of times the flywheel was picked up (23 and 17, respectively), times the average number of flywheels milled per day (70).

Post-Intervention Evaluation: To reduce the amount of weight handled by the flywheel milling operators, and to increase production rates the flywheel milling job, as described in the pre-intervention section above, was divided into two milling cells. Also, instead of two flywheel castings for the left and right half (gear and sprocket sides) as in the initial NIOSH survey, there is one master flywheel casting weighing 17.5 lbs for the FL flywheel, and 13.5 lbs for the XL flywheel. From the FL and XL master castings, the left and right side of the flywheels are milled. In the first flywheel milling cell (Figure 1), 13 steps were required to complete the workcycle. The number of flywheels milled per 8-hour day was approximately 84, this represents 17,472 lbs handled per day for the FL flywheel, and 13,759 lbs for the XL flywheel. In the second flywheel milling cell (Figure 2), 9 steps were required to complete the workcycle. The number of flywheels milled per day for this cell also was approximately 84, representing 12,096 lbs for the FL, and 9,526 lbs for the XL flywheel handled per day. Because of the short cycle time for the second milling cell, the worker on this job also worked on the flywheel balancing job. Table 2 summarizes the material handling results of the flywheel milling job before and after ergonomic interventions.

Hand-Arm Vibration Exposure

Pre-intervention evaluation: In addition to the potential overexertion injuries for manual handling of the flywheels in the milling cells, another concern was excess hand-arm vibration exposure from the use of a hand-held grinder to remove metal burrs from the flywheel. It was determined that approximately 20% of the job cycle was used for removing metal burrs. As noted in Table 1, recommendations were provided to reduce vibration exposure and improve job efficiency. Figure 3 shows a worker using the hand-held grinder.

Post-intervention evaluation: Vibration exposure was virtually eliminated with the purchase of a customized metal deburring machine. This machine was designed according to specifications from engineers and workers performing the job, and would automatically remove burrs with grinding media (stones) inside the unit (see Figure 4). The installation of this unit in the flywheel milling cell resulted in over a 90% reduction in hand-held grinders and a reduction from 20% of the work cycle to less than 1% (occasional touch up) for hand-held grinding operations. The deburring machine allowed the worker to move on to other work elements while this job was done, thus making the job more efficient, and reducing potential hazardous vibration exposure.

The cost of the deburring machine was over 200,000 dollars. To justify the costs over hand-held grinding, the company has established an evaluation program which incorporates the goal of sound engineering and production principles with ergonomic design. Table 3 lists the steps in which decisions and actions of plant personnel accomplished its goal as applied to the deburring machine.

Truing (Assembly and Centering)

Pre-Intervention Truing Evaluation: After milling, the flywheels are assembled together with connecting rods, bearings, and a crank pin. A marriage press is used to sandwich the parts into one unit. The flywheel unit is then "trued." After mounting the flywheel on a fixture, the flywheel unit is manually rotated, using a centering gauge to detect misalignment, the unit is struck using a 5-pound brass-head hammer held by the worker (see Figure 5). Depending on the amount of straightening

Health Hazard Evaluation Report No 91-0208-2422

necessary, the initial impact of the brass-head hammer can be as high as 92,000 lbs. Impact forces are reduced as the flywheel is straightened to specifications. The repeated forces needed to straighten the unit were somewhat traumatizing to the workers and they needed to be rotated from this job ever 2 to 4 hours. Engineers and workers were working on how to reduce exposure to this job when NIOSH researchers arrived during the initial visit in January 1990. NIOSH researchers agreed that the job needed to be changed to reduce the impact force stressors to the upper limbs.

Post-intervention Truing Evaluation: Recommendations to reduce exposure to this job (called the "hammer slammer" job by workers) resulted in the use of a 40 ton press that was modified, based on plant ergonomic committee input, for truing the flywheels (see Figure 6). The press completely eliminated the need for the brass hammers, thus eliminating mechanical trauma to the upper limbs from this task.

Biomechanical analysis of this job showed that there is a moderate risk for lifting the 34 lb. flywheel assembly from the tote bin to the Hess press machine (Figures 7 and 8), Table 4, and from lifting the flywheel from the worktable to the tote bin (Figures 9, and 10, Table 5). Using the NIOSH formula for manual lifting, it was determined that the weight should be no more than approximately 20 lbs for the majority of workers to safely perform this job. As shown in Tables 4 and 5, the two biggest factors which reduced the amount of weight that can be safely handled were the horizontal distance (from the worker's spine to where the hands lift the flywheel), and the amount of upper to lower body asymmetry (i.e., twisting) while the worker handles the flywheel. Because of this, the "safe" lifting limit changes at the origin of the lift, and the destination, 27.2 lbs, and 20.1 lbs respectively (Table 4), and from the worktable, to the tote bin, 28.9 lbs and 19.7 lbs. Therefore, it is prudent to take the lowest weight that can safely be handled from the beginning to the end of the work cycle.

Balancing

Pre-intervention Balancing Evaluation: After the flywheel unit is trued, the next step is balancing. This process involves manually picking up the flywheel from a cart, and placing it in a cradle in the balancing machine (see Figure 11). The flywheel connecting rods are attached to balancing arms which rotate the flywheel at high speeds. Balance sensors relay a profile of the units balance characteristics to a computer which determines where the holes are to be drilled. After the holes are drilled the flywheel unit is rotated once more for a final balance check. The flywheel is manually picked up from the balancing machine, and placed in a cart. The process is then repeated.

Post-intervention Balancing Evaluation: A similar procedure for balancing the flywheels is performed using the sensors and computer. However, because the flywheel unit is heavy, an overhead hoist mounted on an x-y trolley (gantry hoist) was used to lift the unit and place it in the balancing unit cradle (see Figure 12). Balancing is performed by the computer, and the gantry hoist is used once more to put the finished part back in the cart. Using the 1991 NIOSH lifting formula, it was determined that workers performing this job were occasionally at risk for back injury when manually handling the FL flywheel unit. From Figure 11, it was determined that when the flywheel unit is picked up a safe weight is approximately 30 lbs, and when it is placed in the balancing cradle the safe weight is approximately 21 lbs. The difference in safe lifting weights is mainly attributable to the location of the load in relation to the body when it is placed in the balancing cradle. Therefore, the hoist is an excellent engineering control to address this material handling problem. Table 6 summarizes the information used to determine the NIOSH Recommended Weight Limit (RWL) of 30 and 21 lbs.

B. **Rates of Musculoskeletal Disorders**

During the entire 7 years (1987-1993), in the entire production facility there was a total of 555 reports of work-related musculoskeletal disorders, which resulted in 6255 lost workdays and 3385 restricted workdays. This translates to an average of between two and three lost or restricted workdays per worker per year due to work-related musculoskeletal disorders. About 57% of the cases and 65% of the lost or restricted workdays involved the upper extremities (shoulder, elbow, arm, or hand) and 40% of the cases and 33% of the lost or restricted workdays involved the back.

Table 7 shows the yearly incidence rates (the percent of workers reporting a new case each year) of musculoskeletal disorders for the entire production facility, as well as specific rates for the most commonly affected body parts: shoulder, hand/arm, and back. During the four year period (1990-1993) of development and implementation of an ergonomics program at this facility, the incidence rate of musculoskeletal disorders has decreased from 17% to 13%.

Figure 13 shows the increase in the number of hourly workers and the change in incidence rates between 1987 and 1993. Earlier in the 1980's, Harley-Davidson experienced economic difficulties and had laid

Health Hazard Evaluation Report No 91-0208-2422

off some of its workforce. As production began increasing, experienced workers were recalled to work from layoffs. Eventually, in about 1988 or 1989 new and inexperienced workers were hired, which may, in part, explain the rise in cases at that time. Additionally, at that same time, a new nurse was hired who brought new vigilance to the reporting of musculoskeletal disorders.

Table 8 shows the yearly incidence rates for the flywheel milling department between 1987 and 1993. The table shows both the incidence rates for all cases and the incidence rates for only those cases which involved lost or restricted workdays. It also shows the severity rates (the number of lost or restricted workdays per 100 workers) and the median number of lost or restricted workdays per case.

Table 8 shows that in 1989, just prior to the initial request for help, there was a dramatic increase in the incidence rate. Aside from that year, the total incidence rates show no clear pattern of change. The rate of lost or restricted workday cases (the more serious cases), however, has been decreasing from 27% in 1989 to 12.5 % in 1993. Additionally, there has been a decrease in the severity of cases as measured by the number and rate of lost or restricted workdays. This pattern has been erratic, however, because in 1990, prior to the implementation of most of the changes there was a major decline in cases during that one year. This may be explained by a sudden increase of 20% in the size of the department's workforce. Since many of these were new workers, it is possible that during their initial months of employment either they underreported potential musculoskeletal problems or the musculoskeletal disorder developed gradually and did not become symptomatic until the following year.

This follows a pattern seen in other facilities following ergonomic interventions. Initially there may be an increase in the total number of reports of cases as workers become better educated about these problems. However, because of early identification and treatment combined with changes in workstation design, the severity of reported cases declines. Additionally, since many of the ergonomic changes have only recently been introduced, it may still be too early to measure the full impact of the ergonomic program.

V. DISCUSSION

The problem-solving approach used by this company was effective because it involved a team approach of employees, engineers, managers, and medical personnel. This resulted in a participatory approach in which all parties in the flywheel milling and assembly department contributed with their knowledge and experience. Examples of this were demonstrated in the elimination of 5-pound brass hammers by a 40-ton press in the flywheel milling and truing area, elimination of manual transportation of flywheel units to the balancing machine by an overhead hoist system, and elimination of hand held grinders by a spindle deburring machine.

The experience of implementing an ergonomics program at Harley-Davidson exemplifies several general issues about what it takes to sustain a successful ergonomics intervention effort. The first lesson was that problem solving usually includes a series of steps rather than one leap from the problem to a solution. Depending on the training and resources of the company, this process can be immediate or take months. In addition, resources needed to do the job can be nominal or very costly. Examples of the two extremes in this study are: raising the drill press to eliminate stooping while loading flywheels into its fixture (no costs), and purchasing a customized spindle deburring machine (over 200,000 dollars).

Another lesson was that successful ergonomic programs need to be sustained because of the dynamic nature of today's business and production environment. Although a variety of approaches can achieve this, the company found that outside experts were helpful in assessing ergonomic changes, providing stimulus, and acted as a catalyst to move things along. The outside expert proved to be effective on the planning side of the equation, so that ergonomic factors could be engineered into the machines and processes prior to operation. Retrofitting machinery can be very costly compared to engineering ergonomics into original machine design.

The third lesson was understanding the importance of the front line supervisor who serves as the communicator between management and the production worker. The front line supervisor can make or break an ergonomics program. The supervisor provides a supportive environment for worker ideas, and enhances their concepts into practical applications using sound engineering principles. The front line supervisor also needs to effectively communicate with upper management to present needs in a systematic way, and secure resources to get the job done right.

The goal of an effective ergonomic intervention effort is to eliminate the job hazards. At Harley-Davidson this was accomplished through a process where there was commitment from top management to provide resources to manufacture flywheels better and more safely, from company engineers to select the most cost-effective equipment available, and from workers to be involved in every aspect of the equipment from selection to custom design.

Health Hazard Evaluation Report No 91-0208-2422

Because the process of ergonomic changes involved the employees as well as management, and because the employees were involved in all phases of the ergonomics process, it is believed that Harley-Davidson should be used as a model for other industries which have high morbidity from poorly designed jobs.

VI. RECOMMENDATIONS

A. Engineering Controls

Flywheel Milling

1. While the amount of flywheel material has been reduced significantly (from over 28,175 lbs to 17,472 lbs for FL flywheels), during this intervention evaluation, production has increased 7 to 10 % per year. With the combination of higher production, and incentive pay for working above 100%, workers will likely increase the total amount of flywheel material handled over time. Because of this, it is suggested that the ergonomic team think about how to further optimize flywheel throughput with the least amount of manual handling. Some possible ways of accomplishing this include: gravity conveyers, optimal positioning of worktables and machines to reduce travel distances, and orienting machines so that flywheels are positioned at chest level, and close to the body.

2. Plastic separators to separate flywheels in tote carts should be readily available. Flywheel separators should be used for both incoming and outgoing stock. Make sure there are enough plastic separators in the system for the suppliers. Consider bar codes for the separators, and encourage the operators to use the new light pen system located near their workstations to electronically scan the bar code and call for more separators when they get low.

3. The workers should keep the work area as clear as possible. All worktables in the milling area should have purpose, and be positioned so they help with throughput.

4. Rubber matting should be kept in good repair, and be replaced periodically to maintain good cushion and support for the worker.

Health Hazard Evaluation Report No 91-0208-2422

Assembly and Truing Flywheels

1. Install an overhead hoist for moving the flywheel to and from the marriage press and 40-ton Hess press during truing operations. Because of overhead barriers, consider a counter balancing device and articulating arm to overcome these barriers.

2. There is excess manual material handling from moving the flywheel unit back and forth between the Hess press and an adjacent work table used for additional measurements. Consider integrating the two operations. If this is not possible, consider installing a swing arm with this measuring device attached. The swing arm can be pivoted next to the Hess press and the flywheel unit can be moved easier.

Flywheel Balancing

1. Install an adjustable height wooden workbench to optimize biomechanical leverage for reaming holes in flywheels. The current workbench is approximately 5-6 inches too high for the operator who is approximately 6'4" tall.

B. Work Practices

Flywheel Milling

1. Workers should be trained to organize their work station to optimize movement and function. This can be done when tote carts, and tables, can be moved into position by the worker. This is especially helpful, when the same machines are operated by different workers.

Flywheel Balancing

1. Encourage operators to use a hoist, especially when placing and removing the flywheel unit from one balancing machine. Machinery and palm button controls require operator to over reach when manually putting flywheel unit into balancing cradle. Because of this, education and training of workers in using the hoist is recommended.

C. Organizational

1. Develop a written ergonomics program based on the approach used in Department 909, and consider using personnel from this

Health Hazard Evaluation Report No 91-0208-2422

    Videotaping the workers and providing narratives to this process may make it cost effective for the company. The videotape can be used as an orientation for new employees, and for other departments as a place to begin their own program.

2. Consider sit/stand chairs to offer temporary relief from standing between work cycles when machines are performing work in this department.

3. Worker designed caliper set holder in the flywheel milling cell #1 should be showcased as an example of good ergonomic design by the worker and the tool department. The holder serves the worker in making it easy and convenient to access measuring tools for quality control.

D. Medical Surveillance

Develop a medical surveillance program which monitors musculoskeletal disorders in the plant. The ergonomic program and training of plant personnel have raised awareness of job hazards and early reporting of musculoskeletal discomfort is expected. Because of the dynamic nature of manufacturing in this plant, job hazards may vary depending on production demands, quality of parts, and maintenance of machines and tools. Early detection of problems will complete the communication cycle between workers and management to avoid more serious musculoskeletal disorders and to develop priorities for where ergonomic intervention efforts should be focussed.

E. Other

Air hoses used to remove excess oil from the parts should be used with care. When the nozzle is held too close to the flywheel to remove the oil, it can generate oil aerosols which may cause both acute and chronic health problems.

Health Hazard Evaluation Report No 91-0208-2422

VII. REFERENCES

1. **National Institute for Occupational Safety and Health:** Health Hazard Evaluation Report No. HETA 90-134-2064, Harley-Davidson, Inc., Milwaukee, Wisconsin, by J.D. McGlothlin, R.A. Rinsky, and L.J. Fine. Washington, D.C.: Government Printing Office, 1990.

2. **Conn, H.R.:** Tenosynovitis. *Ohio State Med. J. 27*:713-716 (1931).

3. **Pozner, H.:** A Report on a Series of Cases on Simple Acute Tenosynovitis. *J. Royal Army Medical Corps 78*:142 (1942).

4. **Hymovich, L., Lindholm, M.:** Hand, Wrist, and Forearm Injuries. *J. Occup. Med. 8*:575-577 (1966).

5. **National Institute for Occupational Safety and Health:** *Health Hazard Evaluation and Technical Assistance Report No. TA 76-93* by C.L. Wasserman, and D. Badger. Washington, D.C.: Government Printing Office, 1977.

6. **Anderson, J.A.D.:** System of Job Analysis for Use in Studying Rheumatic Complaints in Industrial Workers. *Ann. Rheum. Dis. 31*:226 (1972).

7. **Hadler, N.:** Hand Structure and Function in an Industrial Setting. *Arth. and Rheum. 21*:210-220 (1978).

8. **Drury C.D., Wich, J.:** Ergonomic Applications in the Shoe Industry. In: Proceedings Intl. Conf. Occup. Ergonomics, Toronto, May 7-9, 1984. pp. 489-493.

9. **Cannon, L.:** Personal and Occupational Factors Associated with Carpal Tunnel Syndrome. *J. Occup. Med. 23(4)*:225-258 (1981).

10. **Armstrong, T.J., Foulke, J.A., Bradley, J.S., Goldstein, S.A.:** Investigation of Cumulative Trauma Disorders in a Poultry Processing Plant. *Am. Ind. Hyg. Assoc. J. 43*:103-106 (1982).

11. **Silverstein, B.A.:** "The Prevalence of Upper Extremity Cumulative Trauma Disorders in Industry." Ph.D. Dissertation, University of Michigan, 1985.

12. **Cummings, J., Maizlish, N., Rudolph, M.D., Dervin, K., and Ervin, CA:** Occupational Disease Surveillance: Carpal Tunnel Syndrome. *Morbidity and Mortality Weekly Report July 21, 1989*. pp. 485-489.

13. **Bigos, S.J., Spenger, D.M., Martin, N.A., Zeh, J., Fisher, L., Machemson, A., and Wang, M.H.:** Back Injuries in Industry: A Retrospective Study. II. Injury Factors. *Spine 11*:246-251 (1986a).

14. **Frymoyer, J.W., and Cats-Baril, W.:** Predictors of Low Back Pain Disability. *Clin. Ortho. and Rel. Res. 221*:89-98 (1987).

15. **Magora, A.:** Investigation of the Relation Between Low Back Pain and Occupation. *Ind. Med. Surg. 41*:5-9 (1972).

16. **U.S. Department of Labor, Bureau of Labor Statistics:** *Back Injuries Associated with Lifting*. Bulletin 2144, August 1982.

17. **Chaffin, D.B., and Park, K.S.:** A Longitudinal Study of Low-Back Pain as Associated with Occupational Weight Lifting Factors. *Am. Ind. Hyg. Assoc. J. 34*:513-525 (1973).

18. **Liles, D.H., Dievanyagam, S., Ayoub, M.M., and Mahajan, P.:** A Job Severity Index for the Evaluation and Control of Lifting Injury. *Human Factors 26*:683-693 (1984).

19. **Burton, A.K., and Sandover, J.:** Back Pain in Grand Prix Drivers: A Found Experiment. *Ergonomics 18*:3-8 (1987).

20. **Bureau of National Affairs, Inc.:** *Occupational Safety and Health Reporter*. July 13, 1988. pp. 516-517.

21. **Svensson, H., and Andersson, G.B.J.:** The Relationship of Low-Back Pain, Work History, Work Environment, and Stress. Spine 14:517-522 (1989).

22. **Deyo, R.A., and Bass, J.E.:** Lifestyle and Low-Back Pain: The Influence of Smoking and Obesity. *Spine 14*:501-506 (1989).

23. **Postacchini, F., Lami, R., and Publiese, O.:** Familial Predisposition to Discogenic Low-Back Pain. *Spine 13*:1403-1406 (1988).

24. **Snook, S.H.:** Approaches to the Control of Back Pain in Industry: Job Design, Job Placement, and Education/Training. Spine: State of the Art Reviews 2:45-59 (1987).

25. **Barnes, R.:** *Motion and Time Study, Design, and Measurement of Work*. New York, N.Y.: John Wiley and Sons, 1972.

26. **Gilbreth, F.B.:** *Motion Study*. Princeton, N.J.: Van Nostrand, 1911.

27. **Armstrong, T.J., and Silverstein, B.A.:** Upper-Extremity Pain in the Workplace-Role of Usage in Casualty. In *Clinical Concepts in Regional Musculoskeletal Illness*. Grune and Stratton, Inc., 1987. pp. 33-354.

28. **McGlothlin, J.D.:** "An Ergonomics Program to Control Work-Related Cumulative Trauma Disorders of the Upper Extremities." Ph.D. Dissertation, University of Michigan, 1988.

29. **Waters T.R., Putz-Anderson V., Garg A., Fine L.J.:** Revised NIOSH equation for the design and evaluation of manual lifting tasks. Ergonomics. Vol. 36, No. 7, 749-776.

Health Hazard Evaluation Report No 91-0208-2422

VIII. AUTHORSHIP AND ACKNOWLEDGEMENTS

Principal Investigators:	James McGlothlin, Ph.D, C.P.E
Research Ergonomist
Engineering Control Technology Branch
Division of Physical Sciences
and Engineering

Sherry Baron, M.D., M.P.H
Ergonomic Coordinator
Hazard Evaluations and Technical
Assistance Branch
Division of Surveillance, Hazard
Evaluations, and Field Studies

IX. DISTRIBUTION AND AVAILABILITY

Copies of this report may be freely reproduced and are not copyrighted.

Copies of this report have been sent to:

1. Harley-Davidson Incorporated
2. United Paper Workers International Union, Local 7209

For the purpose of informing affected employees, copies of this report shall be posted by the employer in a prominent place accessible to the employees for a period of 30 calendar days.

Table 1
Evaluation of Ergonomic Changes
HETA 91-0208-2422 Harley-Davidson Incorporated

Engineering Changes

Initial Recommendation (January 1990)	Result (August 1993)
Reduce the weight of the flywheels by improving die-cast specifications. This will reduce milling time and the amount of weight handled over the workday.	Weight of fly wheels were reduced from nearly 2 lbs by improving die-cast specifications. In addition, only one type of flywheel casting (for the gear and sprocket sides) is shipped to plant, and is milled to specifications. This simplifies the milling process, reduces waste, and multiple handling of flywheels.
Reduce or eliminate exposure to vibration from powered hand grinder. Twenty percent of the work cycle time consists of vibration exposure from this tool.	Customized metal deburring machines were purchased to eliminate over 90% of the exposure from the hand grinding operations. The hand grinder is used less than 1% of the work cycle time (for minor touch up of fly wheel).
Layout of the flywheel milling job is inefficient from a production and material handling perspective. Consider movable flywheel carts and/or gravity conveyors between milling work stations to reduce musculoskeletal stress.	The Flywheel milling cell was reorganized into 2 work cells, reducing the number of machines per cell, and the amount of material handled per worker. For the FL flywheel, this resulted in a 38% reduction in material handling from 28,175 lbs to 17,472 per shift, and a 43% reduction in the number of times the operator needed to handle the flywheel during the milling process. Similar results were documented for the XL flywheel milling process.
Reduce the size of the metal pan that is built around the base of the indexing machine and round the corners to reduce the reach distance to attach the flywheels to the machine.	The indexing machine has been eliminated, and replaced by the another more efficient machine. Physical barriers were considered and designed out of the new machine before it was put into operation.
Install durable rubberized floor matting around flywheel milling cells to reduce lower limb fatigue of workers.	Several types of rubberized floor matting were evaluated for durability, slip-resistance, and comfort by the operators in this department. A selection of rubberized mats were made available for the operators.
Remove all physical barriers that may cause workers to overreach, such as limited toe and leg space where the worker has to reach over barriers to manually position flywheels for processing.	Most physical barriers were eliminated because the worker was part of the workstation redesign process. Toe and leg space were considered when the work cells were redesigned. Machines, such as the drill press, were adjusted up to chest height of worker. This reduced stooping to position the flywheels in the machines.

Table 1 (Cont.)
Evaluation of Ergonomic Changes
HETA 91-0208-2422 Harley-Davidson Incorporated

Work Practices

Initial Recommendation (January 1990)	Final Result (August 1993)
Recommend workers use the "power grip" rather than the "pinch grip." when handling the flywheel. The "pinch grip" requires handling of the flywheel by the fingertips and thumb, resulting in high musculoskeletal forces and fatigue. Use of two hands is also recommended when handling parts to reduce asymmetric biomechanical loading of the limbs and back.	All of the workers in this department received ergonomics training on material handling techniques. When the flywheels were handled at the wheel end, both hands were used, especially when positioning flywheels in or out of the milling machines.
When wheel carts are brought into the flywheel milling cell, they should be brought in with the cart bumper facing away from the traffic area to avoid contact with the worker's shins.	The wheel carts bumpers were retrofitted with tubular steel to reduce mechanical contact with the worker's shins. Several of the wheel carts were also fitted with hinged bumpers that can be manually rotated in the vertical position and out of the worker's way. Workers position the wheel carts close to their work area to reduce distance and material handling.
Operators should avoid overreaching while handling flywheels during milling. Overreaching may result in excess musculoskeletal stress and possibly injury, especially later in the work shift when the worker may become fatigued.	On-site training of workers about biomechanical aspects of work may have increased their awareness to reduce overreaching while performing their job. Redesign of the workstation also helped reduce overreaching by providing leg and toe clearance, and adjusting the height of the workstations to fit the worker.

132

Table 2.
Comparison of Pre- and Post- Interventions of Manual Handling of Flywheel Milling
HETA 91-0208-2422 Harley-Davidson Incorporated

Pre-intervention (January 1990)	Fl-Flywheel	Xl-Flywheel
Premilled Flywheel Weight	19.0	16.0
Average Weight	17.5	14.5
Average Cycle Time	5 minutes	4 minutes
# Flywheels/8-hr shift	70	75
# Steps Moving Flywheel	23	17
# lbs. Moved/8-hr shift	28,175	18,980
Post-Intervention First Flywheel Cell[1] (August 1993)		
Premilled Flywheel Weight	17.5	13.5
Average Flywheel Weight	16.0	12.6
Average Cycle Time	4 minutes	4 minutes
# Flywheels/8-hour shift	84	84
# Steps Moving Flywheel	13	13
# lbs. Moved/8-hour shift	17,472	13,759
Post-Intervention Second Flywheel Cell[2]		
Average Weight Flywheel	16.0	12.6
Average cycle time	1.5 minutes	1.5 minutes
# Flywheels/8-hour day	84[3]	84
# Steps Moving Flywheel	9	9
Average Weight/8-hour day	12,096	9,526

1. Flywheel milling completed by another worker in adjacent cell.
2. Second flywheel cell completes the milling process.
3. Up to 280 flywheels can be milled per 8-hour day. However, this worker also does flywheel balancing job and only keeps pace with the first flywheel milling cell.

Table 3
Steps in the Flywheel Deburring Machine Purchase
HETA 91-0208-2422 Harley-Davidson Incorporated

Steps	Activity	Comments
1.	NIOSH report (January 1990), observes potential problem from hand-arm vibration exposure from hand-held grinder.	Recommends several options to reduce exposure, including a metal finishing machine to remove burrs.
2.	Problem solving team formed by the company.	Team participants: Manufacturing engineer -1, operators -2, maintenance machine repairman -1, supervisor -1, tool designer -1, medical -1, purchasing -1, and facility -1.
3.	Mission Statement formed.	"Deburr flywheels and connecting rods in a manner to decrease musculoskeletal injuries from hand grinders, while improving quality and reducing variability."
4.	Overview of Method.	1. Three vendors quoted project; 2. team ran trials with all three and rated results on a matrix; 3. one vendor received highest quality matrix rating and also received a consensus favorable rating from the team.
5.	Definition of Priorities	1. "What the customer wants." (safety, quality, ergonomics); 2. How the company can meet these requirements: (a) reduce hand grinding > 90%, (b) machine construction, (c) ease of load and unload.
6.	Analysis Methods	1. Trials and analysis; 2. interview of vendors; 3. discussion and review of machine and process details.

Table 3 (cont.)
Steps in the Flywheel Deburring Machine Purchase
HETA 91-0208-2422 Harley-Davidson Incorporated

Steps	Activity	Comments
7.	Justification	1. Safety: (a) eliminate flywheel grinding by 90%, (b) estimated savings from prevented lost-time accidents $53,679 (1987-1991). 2. Quality, same or improved - no loose burrs, reduced variability, complexity, increase throughout. 3. Ergonomic, (a) easy to load and unload, (b) no forward bending, especially with weight out in front of body, (c) both hands available to handle flywheels. 4. Housekeeping and environmental, (a) noise cover, (b) eliminates flying metal.
8.	Delivery and Payback Impacts	(a) headcount - meets planned requirements for future layout and schedule production increases, (b) meets capacity effect with less increased manual time, (c) cycle time less than 2 minutes/flywheel, 20 seconds to load and unload, (d) labor cost savings - none except cost increase avoidance savings with increases in schedule, (e) flexibility - increased, (f) set-up less than 10 minutes, (g) in-process inventory, (h) floor space - more than hand grinders; same for alternatives, (i) overhead - increased. annual usage cost saved (grinders and bits $1,730, new process annual costs $7,848)
9.	Employee modification recommendations	(a) Install insulation covers to reduce noise, (b) install load arms to reduce bending over to manually load flywheels, (c) add rinse cycles to clean flywheels.
10.	Costs	$229,616
11.	Timetable	Delivery (12-23-92), Installation 1-11-93), Implementation (6-14-93).

Table 4
Calculations Using 1991 NIOSH Lifting Formula[1] For Flywheel Truing Machine Operator Task 1
HETA 91-0208-2422 Harley-Davidson Incorporated

Job Analysis Worksheet

Department: 909
Truing Machine Operator
Date: August 30, 1993

Job Title: Pick Up Flywheel from Tote Cart

Step 1. Measure and record task variables

Object Weight (lbs)		Hand Location (in)				Vertical Distance (in)	Asymmetric Angle (degrees)		Freq. Rate		Duration	Object Coupling
		Origin: See Figure 7		Dest.: See Figure 8			Origin	Dest.	lifts /min		Hours	
L(avg)	L(Max.)	H	V	H	V	D	A	A	FM			CM
32	34	15	40	20	50	10	30	0	<.2		<2	Fair

Step 2. Determine the multipliers and compute the Recommended Weight Limits (RWL's)

$$RWL = LC \times HM \times VM \times DM \times AM \times FM \times CM$$

ORIGIN $\quad RWL = 51 \times .67 \times .93 \times 1.0 \times .90 \times .95 \times 1.0 = 27.2$ lbs

DESTINATION $\quad RWL = 51 \times .50 \times .85 \times 1.0 \times 1.0 \times .95 \times 1.0 = 20.1$ lbs

Step 3. Compute the LIFTING INDEX

ORIGIN $\quad$ Lifting index = $\dfrac{\text{Object Weight}}{\text{RWL}}$ = $34/27.2$ = 1.25

DESTINATION $\quad$ Lifting index = $\dfrac{\text{Object Weight}}{\text{RWL}}$ = $34/20.1$ = 1.69

[1] See Appendix A for Calculation for the NIOSH lifting formula.

Table 5
Calculations Using 1991 NIOSH Lifting Formula[1] For Flywheel Truing Machine Operator Task 2
HETA 91-0208-2422 Harley-Davidson Incorporated

Job Analysis Worksheet

Department: 909
Job Title: Truing Machine Operator
Date: August 30, 1993

Job Description: Pick Up Flywheel from Work Table

Step 1. Measure and record task variables

Object Weight (lbs)		Hand Location (in)				Vertical Distance (in)	Asymmetric Angle (degrees)		Freq. Rate	Duration	Object Coupling
		Origin: See Figure 9		Dest.: See Figure 10			Origin	Dest.	lifts /min	Hours	
L(avg)	L(Max.)	H	V	H	V	D	A	A	FM		CM
32	34	15	45	22	35	10	0	20	< .2	< 2	Fair

Step 2. Determine the multipliers and compute the Recommended Weight Limits (RWL's)

RWL = LC × HM × VM × DM × AM × FM × CM

ORIGIN   RWL = 51 × .67 × .89 × 1.0 × 1.0 × .95 × 1.0 = 28.9 lbs

DESTINATION   RWL = 51 × .45 × .96 × 1.0 × .94 × .95 × 1.0 = 19.7 lbs

Step 3. Compute the LIFTING INDEX

ORIGIN   Lifting index = Object Weight / RWL = 34/28.9 = 1.17

DESTINATION   Lifting index = Object Weight / RWL = 34/19.7 = 1.73

[1] See Appendix A for Calculation for the NIOSH lifting formula

Table 6
Calculations using 1991 NIOSH Lifting Formula[1] for Flywheel Assembly Lift for Balancing Job.
HETA 91-0208-2422 Harley Davidson Incorporated

Job Analysis Worksheet

Department: 909
Job Title: Balancer
Date: August 30, 1993

Job Description:
Load Flywheel Unit into Balancing Machine

Step 1. Measure and record task variables

Object Weight (lbs)		Hand Location (in)				Vertical Distance (in)	Asymmetric Angle (degrees)		Freq. Rate	Duration	Object Coupling
		Origin		Dest. See Figure 11			Origin	Dest.	lifts /min	Hours	
L(avg)	L(Max.)	H	V	H	V	D	A	A	FM		CM
32	34	15	40	20	35	5	0	30	<.2	<2	Fair

Step 2. Determine the multipliers and compute the Recommended Weight Limits (RWL's)

$$RWL = LC \times HM \times VM \times DM \times AM \times FM \times CM$$

ORIGIN $RWL = 51 \times .67 \times .93 \times 1.0 \times 1.0 \times .95 \times 1.0 = 30$ lbs

DESTINATION $RWL = 51 \times .50 \times .96 \times 1.0 \times .90 \times .95 \times 1.0 = 21$ lbs

Step 3. Compute the LIFTING INDEX

ORIGIN Lifting index = Object Weight / RWL = 34/30 = 1.13

DESTINATION Lifting index = Object Weight / RWL = 34/21 = 1.62

[1] See Appendix A for Calculation for the NIOSH lifting formula.

**Table 7
Entire Production Facility
Incidence Rates of Musculoskeletal Disorders from OSHA 200 Logs
HETA 91-0208-2422 Harley-Davidson Incorporated**

Year	Total Rate (% of workers)	Shoulder Rate (% of workers)	Hand/Arm Rate (% of workers)	Back Rate (% of workers)
1987	9	3	2	3
1988	10	1	4	4
1989	14	1	5	6
1990	17	5	5	7
1991	17	4	3	7
1992	16	3	4	6
1993	13	3	3	4

**Table 8
Flywheel Department
Trends in Work-Related Musculoskeletal Disorders 1987-1993
HETA 91-0208-2422 Harley-Davidson Incorporated**

Year	Workers	WRMD[1] Incidence Rate (% of Workers)		Lost/Restricted Workdays	
		Total	Lost/Restricted Workday Cases	Number per 100 Workers	Median # per Case[2]
1987	34	17.6	11.8	110	10
1988	34	11.8	8.9	130	13
1989	36	38.9	27.6	610	13
1990	44	20.5	11.5	390	33
1991	43	27.9	18.7	480	21
1992	45	17.8	13.4	560	12
1993	48	20.8	12.5	190	11

[1] Work-related Musculoskeletal Disorders-Includes all neck, upper extremity and back cases.
[2] This includes only those cases that had some lost or restricted workdays.

Figure 1

Flywheel Milling Job #1

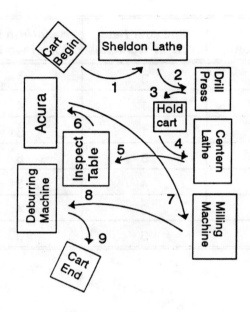

Figure 2

Flywheel Milling Job #2

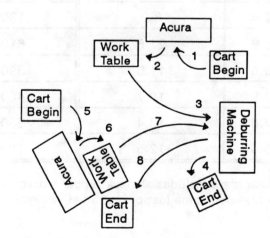

**Figure 3.
Worker in Flywheel Milling job using a hand-held grinder**

Figure 4.
Worker in Flywheel Milling positioning a flywheel on deburring machine robot arm.

Figure 5.
Flywheel truing operations using 5-pound brass-head hammer.

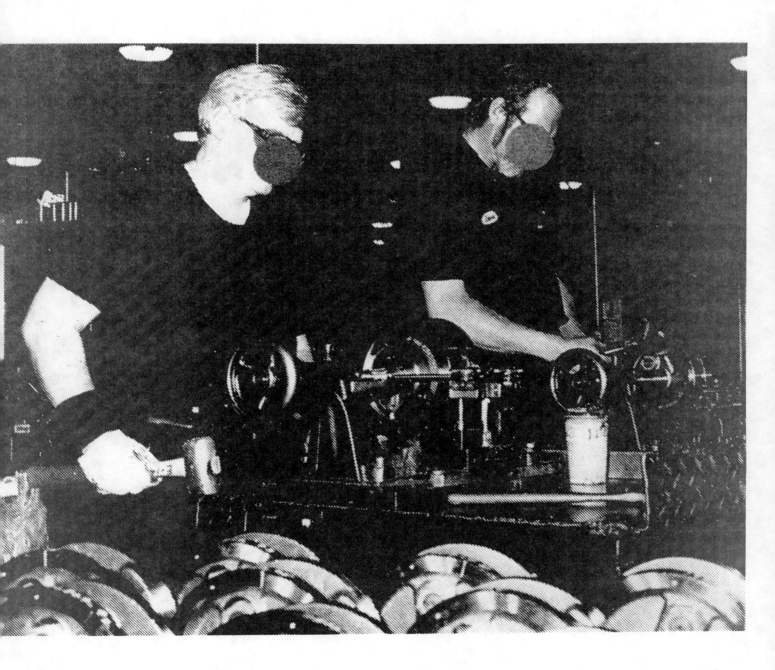

Figure 6.
Flywheel truing operations using a 40 ton press.

Figure 7.
Worker lifting flywheel assembly from tote bin to truing press.

**Figure 8.
Worker lifting flywheel assembly from truing press.**

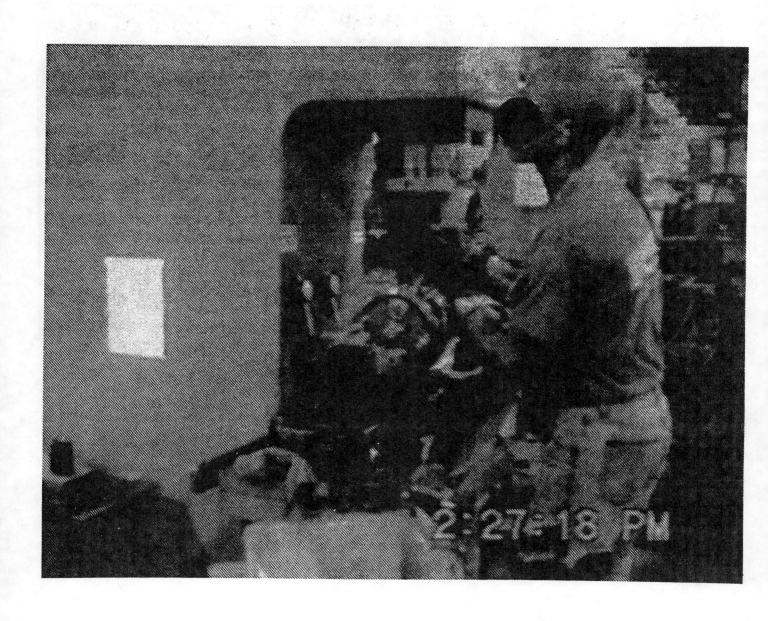

Figure 9.
Worker placing flywheel assembly on worktable (truing area).

Figure 10.
Worker placing flywheel assembly in tote bin (truing area).

**Figure 11.
Worker placing flywheel in balancing machine**

**Figure 12.
Overhead hoist moving flywheel to balancing machine**

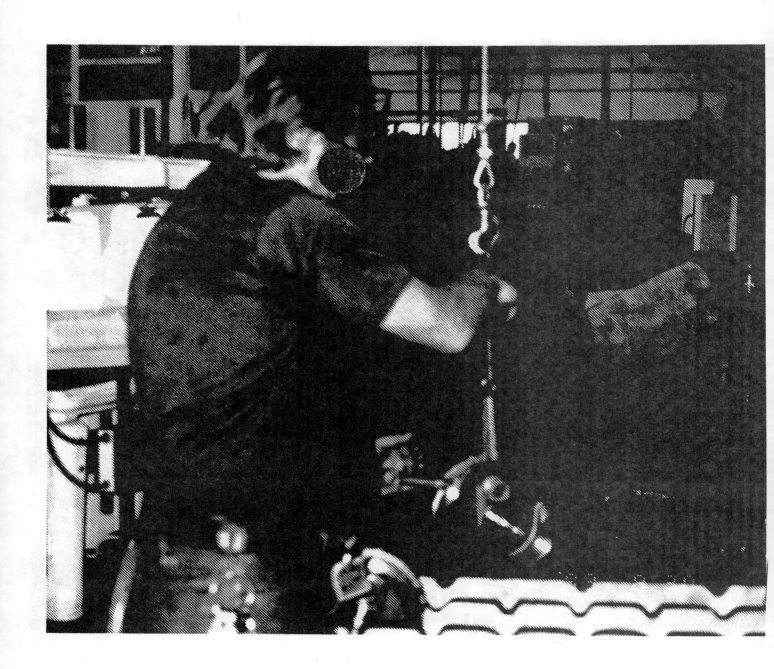

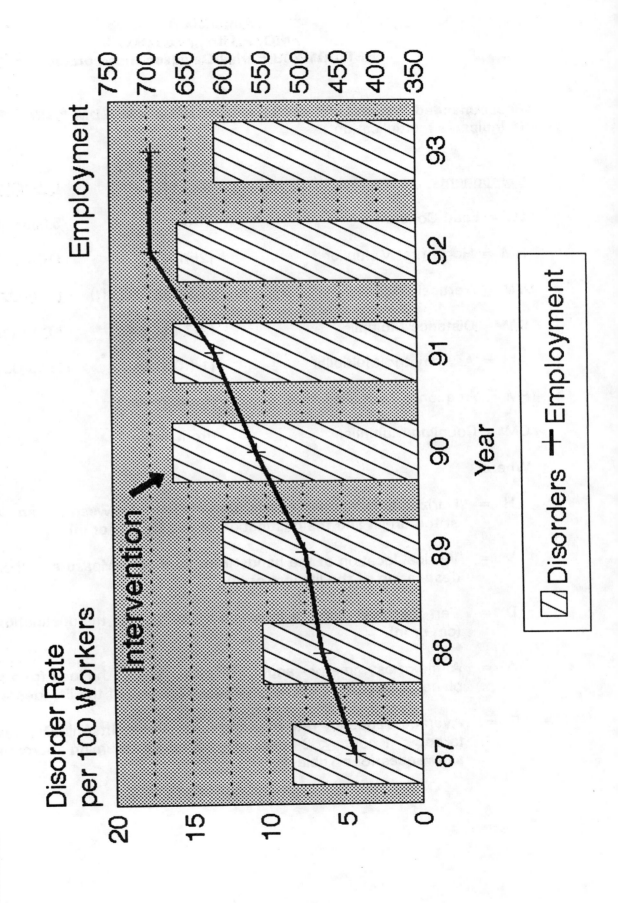

Figure 13
Employment Growth and Rates of Musculoskeletal Disorders

Appendix A
NIOSH Lifting Equation
HETA 91-208 Harley-Davidson Incorporated

Recommended Weight Limit (RWL) = LC * HM * VM * DM * AM * FM * CM
(* indicates multiplication.)

Component	METRIC	U.S. CUSTOMARY
LC = Load Constant	23 kg	51 lbs
HM = Horizontal Multiplier	(25/H)	(10/H)
VM = Vertical Multiplier	(1-(.003|V-75|))	(1-(.0075|V-30|))
DM = Distance Multiplier	(.82+(4.5/D))	(.82+(1.8/D))
AM = Asymmetric Multiplier	(1-(.0032A))	(1-(.0032A))
FM = Frequency Multiplier	(from Table 1)	
CM = Coupling Multiplier	(from Table 2)	

Where:

- H = Horizontal location of hands from midpoint between the ankles. Measure at the origin and the destination of the lift (cm or in).

- V = Vertical location of the hands from the floor. Measure at the origin and destination of the lift (cm or in).

- D = Vertical travel distance between the origin and the destination of the lift (cm or in).

- A = Angle of asymmetry - angular displacement of the load from the sagittal plane. Measure at the origin and destination of the lift (degrees).

- F = Average frequency rate of lifting measured in lifts/min. Duration is defined to be: $\leq$ 1 hour; $\leq$ 2 hours; or $\leq$ 8 hours assuming appropriate recovery allowances (See Table 1).

Appendix A

Table 1
Frequency Multiplier (FM)
NIOSH Lifting Equation

Frequency Lifts/min	Work Duration					
	≤ 1 Hour		≤ 2 Hours		≤ 8 Hours	
	V < 75	V ≥ 75	V < 75	V ≥ 75	V < 75	V ≥ 75
0.2	1.00	1.00	.95	.95	.85	.85
0.5	.97	.97	.92	.92	.81	.81
1	.94	.94	.88	.88	.75	.75
2	.91	.91	.84	.84	.65	.65
3	.88	.88	.79	.79	.55	.55
4	.84	.84	.72	.72	.45	.45
5	.80	.80	.60	.60	.35	.35
6	.75	.75	.50	.50	.27	.27
7	.70	.70	.42	.42	.22	.22
8	.60	.60	.35	.35	.18	.18
9	.52	.52	.30	.30	.00	.15
10	.45	.45	.26	.26	.00	.13
11	.41	.41	.00	.23	.00	.00
12	.37	.37	.00	.21	.00	.00
13	.00	.34	.00	.00	.00	.00
14	.00	.31	.00	.00	.00	.00
15	.00	.28	.00	.00	.00	.00
>15	.00	.00	.00	.00	.00	.00

†Values of V are in cm; 75 cm = 30 in.

Appendix A

Table 1
Frequency Multiplier (FM)
NIOSH Lifting Equation

Frequency	Work Duration					
Lifts/min	≤ 1 Hour		≤ 2 Hours		≤ 8 Hours	
	V < 75	V ≥ 75	V < 75	V ≥ 75	V < 75	V ≥ 75
0.2	1.00	1.00	.95	.95	.85	.85
0.5	.97	.97	.92	.92	.81	.81
1	.94	.94	.88	.88	.75	.75
2	.91	.91	.84	.84	.65	.65
3	.88	.88	.79	.79	.55	.55
4	.84	.84	.72	.72	.45	.45
5	.80	.80	.60	.60	.35	.35
6	.75	.75	.50	.50	.27	.27
7	.70	.70	.42	.42	.22	.22
8	.60	.60	.35	.35	.18	.18
9	.52	.52	.30	.30	.00	.15
10	.45	.45	.26	.26	.00	.13
11	.41	.41	.23	.23	.00	.00
12	.37	.37	.21	.21	.00	.00
13	.00	.34	.00	.00	.00	.00
14	.00	.31	.00	.00	.00	.00
15	.00	.28	.00	.00	.00	.00
≥ 15	.00	.00	.00	.00	.00	.00

Values of V are in inches and 75 in. cm = 30 in.

HETA 90-246-2314
MAY 1993
AMERICAN FUEL CELL AND
  COATED FABRICS COMPANY
MAGNOLIA, ARKANSAS

NIOSH INVESTIGATORS:
GREGORY A. BURR, C.I.H.
YVONNE BOUDREAU, MD
KATHARYN A. GRANT, Ph.D.
DANIEL HABES, M.S.E
MATTHEW K. KLEIN, P.E.

SUMMARY

A request was submitted by the United Rubber Workers (URW) for a National Institute for Occupational Safety and Health (NIOSH) Health Hazard Evaluation (HHE) at the American Fuel Cell and Coated Fabrics Company (Amfuel) plant in Magnolia, Arkansas. Health effects, including "neurotoxic responses, nausea, dermatitis, multiple cancers, narcosis, emotional stress, heat stress, and ergonomic problems," were reported by the union to be occurring among Amfuel workers during the manufacture of coated rubber aircraft fuel cells. An initial survey was conducted on November 14-16, 1990, following the completion of an Occupational Safety and Health Administration (OSHA) safety and health compliance inspection. Two follow-up surveys were conducted by NIOSH investigators in July and August 1991, to measure solvent levels, evaluate heat stress conditions, conduct an ergonomic evaluation, and assess the adequacy of existing ventilation systems.

Ergonomic assessments were performed in the *Fittings, Innerliner, Outerply, Final Inspection, Nylon Spray, and Onion Tank Assembly* areas. A job analysis was performed to assess the repetitiveness of various fuel cell assembly tasks and to document instances of awkward hand, wrist, arm and trunk postures. Manual force requirements were estimated, and exposures to hand/arm vibration were also noted. Medical interviews were conducted and injury and illness records were reviewed.

Personal breathing-zone (PBZ) and general area (GA) air samples were collected for methyl ethyl ketone (MEK) and 1,1,1-trichloroethane, the principal solvents used in assembling and cleaning fuel cells. Concentrations of MEK in the PBZ samples ranged from <10 parts per million (ppm) to 421 ppm, expressed as time-weighted averages (TWAs) over the sampling period. Three of four short-term (15 minutes) PBZ air samples collected for MEK during a ring cleaning operation in the Fittings Department had concentrations which exceeded the NIOSH Short-term Exposure Limit (STEL) of 300 ppm.

Concentrations of 1,1,1-trichloroethane in five short-term (15 to 19 minutes) PBZ air samples collected during the interior cleaning of fuel cells in the Final Finish Department ranged from 293 to 878 ppm. Four of these PBZ air samples had 1,1,1-trichloroethane concentrations which exceeded the NIOSH STEL of 350 ppm. Results from all the PBZ and GA air samples collected for 1,1,1-trichloroethane ranged from <10 ppm to 878 ppm, TWA over the sampling period.

Injury and illness records contained on OSHA injury and illness forms were reviewed for information pertinent to the HHE request. Private medical interviews were conducted with 26 current employees who volunteered to talk about their work-related health concerns. Union representatives provided NIOSH with a list of 30 other current or former employees with work-related health concerns. The most commonly reported health concern was skin rashes. This was attributed by affected employees to "shiny" gum material used in the Innerliner area. The skin rashes were also reported to be worse during the warmer summer months. Those who reported skin rashes also reported that the use of gloves or wearing long sleeves was impractical because it did not allow them to do their job properly. One person interviewed had what she considered a work-related skin rash on an exposed area of her forearm which consisted of several macular red pinpoint-sized areas.

Reported colon and breast cancers occurred in fewer people than would be expected in the general population. There were eight other reported cases of "cancer," but these were not verified by available records and the affected individuals were not available for interview. Other reported health concerns included dizziness, headaches, asthma, eye, nose and throat irritation, sinus congestion, nausea, nervousness, lung disease, thyroid disease, and carpal tunnel syndrome.

The ergonomic recommendations offered in this report include replacing manual cutting shears with powered shears; providing a fixture or tool to remove scrap rubber from the hole-cutting die; re-designing scissors with longer handles, shorter blades, and a self-opening mechanism to reduce the manual stress associated with prolonged and repetitive tool use; changing the height of the work stations to reduce the occurrence of awkward wrist postures and long reaches; providing stools, cushioned floor mats, or raised foot rests; providing additional lighting and (where needed) magnifying glasses to improve visibility; adding rollers to the bottom surface of platforms and racks to allow the operator to transfer molds between surfaces with less force exertion; reducing tool vibration; and modifying tool handles to eliminate conditions which require a pinch grip.

In general, local exhaust ventilation systems were either absent or only partially effective. The company's personal protection program was not effective, evidenced by improper wearing of respirators by employees and the lack of suitable skin protection while handling solvents. The company lacked a confined space entry program and their written respiratory protection program was inadequate. A review of the company's injury and illness records revealed several departments with ergonomic problems such as cumulative trauma disorder (CTD). Specific recommendations for modifying tools, work stations, and work methods were presented to the company.

NIOSH investigators have concluded that multiple health hazards exist at this facility, including overexposures to 1,1,1-trichloroethane and methyl ethyl ketone, an inadequate confined space entry program, numerous ergonomic hazards, and inadequate personal protection. In addition, the NIOSH investigators conclude that both mechanical and chemical trauma to the skin could occur among workers handling organic solvents, rubber adhesives, and rubber stock. Both specific and generalized recommendations have been included in this report to reduce solvent exposures and improve local exhaust ventilation. Recommendations are also included which address ergonomic problems, respirator selection, personal protection, and implementation of a heat stress program.

Keywords: SIC 3069 (Fabricated Rubber Products, Not Elsewhere Classified), methyl ethyl ketone, 1,1,1-trichloroethane, heat stress, ergonomics, confined spaces, respiratory protection, ventilation, personal protective equipment, skin rash, cancer.

Health Hazard Evaluation Report No. 90-246

INTRODUCTION

A request for a National Institute for Occupational Safety and Health (NIOSH) Health Hazard Evaluation (HHE) was submitted by a representative of the United Rubber Workers's (URW) union concerning chemical exposures, heat stress, and ergonomic problems, occurring during the manufacture of coated rubber fuel cells by workers at the American Fuel Cell and Coated Fabrics Company (Amfuel) in Magnolia, Arkansas. Health effects noted in the request included "neurotoxic responses, nausea, dermatitis, multiple cancers, narcosis, emotional and heat stress, ergonomic problems, and pulmonary hemorrhage." An initial survey was conducted by NIOSH investigators on November 14-16, 1990, following the completion of an extensive Occupation Safety and Health Administration (OSHA) safety and health compliance inspection. Two follow-up surveys were conducted by NIOSH investigators in July and August 1991 to measure solvent exposures and evaluate heat stress conditions and the adequacy of existing ventilation systems throughout the facility. In addition, an ergonomic evaluation was conducted in the following departments: *Fittings; Innerliner (including rubber cutting); Outerply; Nylon Spray; Final Inspection; and Onion Tank Assembly.*

Prior to this NIOSH survey, an OSHA heath and safety compliance investigation was conducted at Amfuel between July to September, 1990. Health and safety related citations resulting from this survey were received by the company on November 16, 1990. The OSHA citations initially issued to Amfuel involved an inadequate hazard communication program, inadequate personal protective equipment, and documentation of numerous employee overexposures to methyl ethyl ketone (MEK); methyl chloroform (1,1,1-trichloroethane); methyl isobutyl ketone; lead; morpholine; and carbon disulfide.

BACKGROUND

Formerly owned and operated by the Firestone Tire and Rubber Company, the facility became the American Fuel Cell and Coated Fabrics Company in 1983. The majority of Amfuel employees work producing aircraft fuel cells at one of two plants located in Magnolia. Other production activities (unrelated to fuel cell production) performed at the second Magnolia facility include contract maintenance work on Titan missile components and Mark IV army rocket casings. A smaller plant, located in Monticello, Arkansas, produces transportable 2000 gallon water storage tanks (termed "onion tanks" because of their unique shape when filled with water). The total workforce of all the Amfuel plants was approximately 600 people at the time of this evaluation.

The primary Amfuel products, and the focus of this evaluation, are fuel bladders (also called fuel cells), which are used in military and small commercial aircraft. To maximize fuel storage capacity, these cells typically conform to the shape of the particular aircraft in which they are used. As a result, the shape of the fuel cells may

Health Hazard Evaluation Report No. 90-246

be complex. The assembly time of a single fuel cell may take more than a month due to the manual assembly steps required to produce a final product. The capacity of the fuel cells produced by Amfuel range from several hundred to several thousand gallons. Table 1 lists the departments which were included in this NIOSH evaluation and the type of assessments performed in each area.

The manufacture of fuel cells for aircraft is a labor intensive process which involves the assembly, by hand, of thin layers of material (either fabric or rubber) over mandrels. Mandrels are precision forms which replicate the shape of the fuel cell being produced. Produced off-site by another company, these forms are constructed from paper maché, cardboard, and plaster. Each mandrel is used only once.

Table 1 - NIOSH EVALUATION

Department	Survey Assessments Performed
Rubber Cutting	Ergonomic
Innerliner	Ergonomic, Solvent Exposures
Outerply	Ergonomic, Solvent Exposures
Fittings	Ergonomic, Heat Stress
Cement "House"	Solvent Exposures
Face Coating	Ergonomic, Solvent Exposures, Heat Stress
Heat Treating	Heat Stress
Cement Spray	Solvent Exposures
Nylon Spray	Solvent Exposures
Final Finish	Ergonomic, Solvent Exposures, Heat Stress, Confined Spaces
Onion Tank Assembly	Ergonomic

Large sheets of stock rubber and coated fabric are cut to their desired shape in the *Rubber Cutting* department. The build-up of the multiple layers of rubber and fabric occurs in either the *Innerliner* or *Outerply* Departments. The layers of rubber and fabric are assembled using adhesives (referred to as "cements" by Amfuel workers) which contain solvents (primarily methyl ethyl ketone [MEK]) and additives (such as carbon black). The adhesives are manually applied by brush or roller to the hand-held sheets of material. Metal fittings, loops, and other specialty connections may be attached to the cells in these departments.[a] Nylon and latex coatings are sprayed on the cells during various stages of production.

Following the *Innerliner* and *Outerply* Departments, the assembled fuel cells are autoclaved (a rubber curing process similar to vulcanization). Prior to autoclaving, the

---

[a] The employees located in the adjacent *Fittings* Department prepare sub-assemblies consisting of metal hardware (rings, bolts, etc.) which are bonded to rubber collars. These sub-assemblies are then shipped to the *Innerliner* Department for attachment to the appropriate fuel cell.

Health Hazard Evaluation Report No. 90-246

cardboard mandrel is softened by soaking the fuel cell in a large tank of heated water and then removed. Following autoclaving, all of the fuel cells, as part of a strict quality control program, are leak tested by inflating them with air to identify any structural defects. In addition, depending on the military specifications required for a particular fuel cell, additional leak checks, using jet fuel, may be performed.

The last stage prior to shipping is the *Final Finish* Department where all the cells receive a final inspection, repair (if needed), and final cleaning of both interior and exterior surfaces. Typically one or two inspectors (depending on the size of the fuel cell) inspect and clean a cell. Cleaning solvents used in this department include MEK (for the exterior surfaces) and 1,1,1-trichloroethane (for the interior portions of the cell). Since the fuel cells have little structural rigidity to them once the mandrel is removed, the workers suspend the cells from elevated work tables using rigging consisting of small ropes attached from overhead metal bars to loops on the exterior surface of the fuel cell. This arrangement allows the workers to inspect both the exterior and interior surfaces. Depending on the size of the cell, the employees may be required to work completely inside the cell.

EVALUATION CRITERIA

GENERAL

As a guide to the evaluation of the hazards posed by workplace exposures, NIOSH field staff employ environmental evaluation criteria for the assessment of a number of chemical and physical agents. These criteria are intended to suggest limits of exposure to which most workers may be exposed up to 10 hours per day, 40 hours per week for a working lifetime without experiencing adverse health effects. It is, however, important to note that not all workers will be protected from adverse health effects even though their exposures are maintained below these limits. A small percentage may experience adverse health effects because of individual susceptibility, a pre-existing medical condition, and/or a hypersensitivity (allergy). In addition, some hazardous substances may act in combination with other workplace exposures, the general environment, or with medications or personal habits of the worker to produce health effects even if the occupational exposures are controlled at the limit set by the criterion. These combined effects are often not considered in the evaluation criteria. Also, some substances are absorbed by direct contact with the skin and mucous membranes, and thus potentially increase the overall exposure. Finally, evaluation criteria may change over the years as new information on the toxic effects of an agent become available.

The primary sources of environmental evaluation criteria for the workplace are the following: 1) NIOSH Recommended Exposure Limits (RELs),[1] 2) the American Conference of Governmental Industrial Hygienists' (ACGIH) Threshold Limit

Health Hazard Evaluation Report No. 90-246

Values (TLVs),[2] and 3) the U.S. Department of Labor, Occupational Safety and Health Administration (OSHA) Permissible Exposure Limits (PELs).[3] The OSHA PELs may be required to take into account the feasibility of controlling exposures in various industries where the agents are used; the NIOSH RELs, by contrast, are based primarily on concerns relating to the prevention of occupational disease. In evaluating the exposure concentrations and the recommendations for reducing these concentrations found in this report, it should be noted that the lowest exposure criteria was used; however, industry is legally required to meet those limits specified by the OSHA standard.

A time-weighted average (TWA) exposure refers to the average airborne concentration of a substance during a normal 8- to 10-hour workday. Some substances have recommended short-term exposure limits (STELs) or ceiling values which are intended to supplement the TWA where there are recognized toxic effects from high short-term exposures.

METHYL CHLOROFORM, METHYL ETHYL KETONE, ETHANOL

Table 2 summarizes toxicity and permissible exposure information on methyl chloroform, MEK, and ethanol. Chloroprene, although listed as an ingredient in the latex spraying operation, was not detected in any of the personal breathing-zone air samples collected for this material. As a result, no toxicity and exposure data has been included for this chemical.

ERGONOMICS

Cumulative trauma disorder (CTD) of the musculoskeletal system is an umbrella term which describes a number of injuries affecting the tendons, tendon sheaths, muscles, and nerves of the upper extremities. Common CTDs include tendinitis, synovitis, tenosynovitis, bursitis, ganglionic cysts, strains, DeQuervain's disease, and carpal tunnel syndrome (CTS).

In recent years, the link between CTDs and occupation has gained increasing attention. In 1990, CTDs were responsible for more than half of all occupational illnesses reported to the Bureau of Labor Statistics.[4] Studies have shown that CTDs can be precipitated or aggravated by activities that require repeated or stereotyped movements, large applications of force in awkward postures, or exposure to hand/arm vibration.[5-7] Postures often associated with upper extremity (UE) CTDs are extension, flexion, and ulnar and radial deviation of the wrist, open-hand pinching, twisting movements of the wrist and elbow, and reaching over shoulder height. Activities associated with UE CTDs are frequently observed in many manufacturing and assembly jobs in industry. Occupations associated with a high incidence of CTDs include electronic components assembly, garment manufacturing, small appliance manufacturing and assembly, and meat and poultry processing.[8-10]

HEAT STRESS

There are a number of heat stress guidelines that are available to protect against heat-related illnesses. These include, but are not limited to, the wet bulb globe temperature (WBGT), Belding-Hatch heat stress index (HSI), and effective temperature (ET).[11-13] The underlying objective of these guidelines is to prevent a worker's core body temperature from rising excessively. Many of the available heat stress guidelines, including those proposed by NIOSH and the ACGIH, use a maximum core body temperature of 38°C as the basis for the environmental criterion.[14,15]

Wet Bulb Globe Temperature (WBGT) Index

Both NIOSH and ACGIH recommend the use of the WBGT index to measure environmental factors because of its simplicity and suitability in regards to heat stress. The WBGT index takes into account environmental conditions such as air velocity, vapor pressure due to atmospheric water vapor (humidity), radiant heat, and air temperature, and is expressed in terms of degrees Fahrenheit (or degrees Celsius). Measurement of WBGT is accomplished using an ordinary dry bulb temperature (DB), a natural (unaspirated) wet bulb temperature (WB), and a black globe temperature (GT) as follows:

$$WBGT_{in} = 0.7 (WB) + 0.3 (GT)$$
for inside or outside without solar load,
OR
$$WBGT_{out} = 0.7 (WB) + 0.2 (GT) + 0.1 (DB)$$
for outside with solar load.

Originally, NIOSH defined excessively hot environmental conditions as any combination of air temperature, humidity, radiation, and air velocity that produced an average WBGT of 79°F (26°C) for unprotected workers.[16] However, in the revised criteria for occupational exposure to hot environments, NIOSH provides diagrams (see Figure 1) showing work-rest cycles and metabolic heat versus WBGT exposures which should not be exceeded.[14]

Similarly, ACGIH recommends TLVs® for environmental heat exposure permissible for different work-rest regimens and work loads.[15] The NIOSH REL and ACGIH TLV criteria assume that the workers are heat acclimatized, are fully clothed in summer-weight clothing, are physically fit, have good nutrition, and have adequate salt and water intake. Additionally, they should not have a pre-existing medical condition that may impair the body's thermoregulatory mechanisms. For example, alcohol use

Health Hazard Evaluation Report No. 90-246

and certain therapeutic and social drugs may interfere with the body's ability to tolerate heat. Modifications of the NIOSH and ACGIH evaluation criteria should be made if the worker or conditions do not meet the previously defined assumptions. Modifications which have been suggested are shown in Table 3.

Selection of a protective NIOSH WBGT exposure limit from Figure 1 is contingent upon identifying the appropriate work-rest schedule and the metabolic heat produced by the work. The work-rest schedule is characterized by estimating the amount of time the employees work to the nearest 25%. The most accurate assessment of metabolic heat production is to actually measure it via calorimetry. However, this is impractical in industrial work settings. An estimate of the metabolic heat load can be accomplished by dividing the work activity into component tasks and adding the time-weighted energy rates for each component. Because of the error associated with estimating metabolic heat, NIOSH recommends using the upper value of the energy expenditure range to allow a margin of safety. Table 4 presents the metabolic estimate for an employee working in the *Final Finish* Department as an example of this technique.[14]

Table 3 - Heat Stress Criteria

1. Unacclimatized or physically unconditioned - subtract 4°F (2°C) from the permissible WBGT value for acclimatized workers.
2. Increased air velocity (above 1.5 meters per second or 300 feet per minute) - add 4°F (2°C). This adjustment can not be used for air temperatures in excess of 90-95°F (32-35°C). This correction does not apply if impervious clothing is worn.
3. Impervious clothing which interferes with evaporation:
   a. Body armor, impermeable jackets - subtract 4°F (2°C)
   b. Raincoats, turnout coats, full-length coats - subtract 7°F (4°C).
   c. Fully encapsulated suits - subtract 9°F (5°C).
4. Obese or elderly - subtract 2-4°F (1-2°C).
5. Female - subtract 1.8°F (1°C). This adjustment, which is based on a supposedly lower sweat rate for females, is questionable since the thermoregulatory differences between the sexes in groups that normally work in hot environments are complex.[17] Seasonal and work rate considerations enter into determining which sex is better adapted to work in hot environments.[18]

Aural Temperature Measurements

As an evaluation technique, the WBGT method is at best only an imprecise indicator of the heat load experienced by a worker. Assumptions must be made regarding the worker's degree of acclimatization and physical fitness. The NIOSH heat stress REL must be adjusted for weight, clothing, work rates, and metabolic heat production. In addition, these heat stress indices may not appropriate for situations where clothing that inhibits or prevents evaporative heat loss (e.g., personal protective equipment) is worn.

Direct measurements of core body temperature typically entail unacceptably invasive techniques (rectal temperature) or require strictly controlled procedures (oral temperature). Commercially available personal heat stress monitors have been developed that are capable of monitoring workers on a continuous basis through a variety of techniques, including ear canal temperature. These monitors generally offer data logging capability, as well as alarm functions for alerting workers when pre-set limits are exceeded. During this survey NIOSH industrial hygienists conducted a limited evaluation using a personal heat stress monitor that utilized the ear-canal temperature technique. It should be noted that the accuracy and precision of this monitor has not been evaluated by NIOSH.

CONFINED SPACES

The NIOSH definition of a confined space is an area which by design has *limited* openings for entry and exit, *unfavorable natural ventilation* which could contain (or produce) dangerous air contaminants, and which is *not intended for continuous employee occupancy*.[19] The NIOSH criteria for working in confined spaces further classifies confined spaces based upon the characteristics such as oxygen level, flammability, and toxicity. As shown in Table 5, if any of the hazards present a situation which is immediately dangerous to life or health (IDLH), the confined space is designated **Class A**. A **Class B** confined space has the potential for causing injury and/or illness but is not IDLH. A **Class C** space would be one in which the hazard potential would not require any special modification of the work procedure. Table 6 lists the items which should be considered before entering any confined space.

EVALUATION DESIGN

AIR MONITORING

Personal breathing-zone (PBZ) and general area (GA) air samples were collected on August 20-21, 1991, for methyl ethyl ketone (MEK), methyl chloroform (1,1,1-trichloroethane), chloroprene, and ethanol. The methods used for collection and analysis of these materials are summarized in Table 7.

VENTILATION ASSESSMENT

A qualitative "wall to wall" ventilation assessment was performed in the fuel cell production areas. This inspection included a visual observation of existing local exhaust ventilation (LEV) controls as well as a review of work practices.

HEAT STRESS

Environmental measurements were obtained using a Reuter Stokes RSS 211D Wibget® heat stress meter manufactured by Reuter Stokes, Canada. This direct reading instrument is capable of monitoring dry bulb, natural (unaspirated) wet bulb, and black globe temperatures in the range between 32° and 200°F, with an accuracy of $\pm$ 0.5-1.0°F. This meter also computes the indoor and outdoor WBGT indices in the

Health Hazard Evaluation Report No. 90-246

range between 32° and 200°F. Measurements were collected about four feet from the floor after the meter was allowed to stabilize.

In addition to the environmental heat stress measurements obtained with the Reuter Stokes Wibget®, a Quest QuesTemp°II® (Quest Electronics, Oconomowoc, WI) personal heat stress monitor was used to measure aural (ear) temperature of an employee working within a fuel cell in the *Final Finish* Department. This device electronically measures temperature in the ear canal. The difference between the ear and body temperatures is compensated for by calibrating the unit directly to the worker's oral temperature. A small sensor, which is placed in the ear canal via an earplug, monitors changes in the body's temperature and will alarm if the level exceeds a pre-set limit (factory set at 38°C, adjustable up to 39°C). The monitor also continuously logs body temperature for subsequent evaluation, such as assessing the heat stress incurred from specific tasks. The ear mold containing the plug and sensor is equipped with a second temperature sensor, which monitors the worker's environment, and a small speaker used for an audible alert. It should be noted that the secondary sensor provides only an estimate of ambient temperature because the values may be affected by its proximity to the worker's head.

According to the manufacturer, this device provides a direct estimate of heat stress on a worker. Because the ear canal borders the hypothalamus (the body's temperature regulator at the base of the brain), if the ear canal is isolated from the outside environment, the sensor will track the temperature of the hypothalamus. In addition to temperature monitoring by the QuesTemp II, standard WBGT readings were measured every five minutes in the immediate work area of this employee.

MEDICAL

Union representatives provided NIOSH with a list of names and work-related health concerns of current and past workers. Current workers were informed by union representatives of the NIOSH visit prior to our arrival. A list was compiled of those current employees who wanted to talk with us about their work-related health concerns and these employees were interviewed privately. In addition, the supervisors of the *Outerply, Repair, Final Finish, Spray Room*, and *Innerliner* departments were interviewed in order to assess the magnitude of health concerns voiced by employees to their supervisors. All interviews were done on a voluntary basis and no one refused to be interviewed. OSHA 200 logs were reviewed for information pertinent to the hazard request.

ERGONOMICS

The first objective of the ergonomic evaluation was to identify biomechanical risk factors for upper extremity (UE) CTDs in jobs performed at these facilities. A second

Health Hazard Evaluation Report No. 90-246

objective was to develop recommendations to eliminate or reduce the hazards identified in these jobs. Jobs performed in the facility during the site visit were observed and videotaped. A total of eight operations were included in the evaluation. Additional information, such as the number of workers employed in each job, the types of tools used, the work station dimensions, and the force requirements of certain tasks, was also collected.

A job analysis was performed to assess the repetitiveness of each task and to document instances of awkward hand, wrist, arm, and trunk postures. Manual force requirements were estimated, and exposures to hand/arm vibration were also noted, although no direct measurements of hand-transmitted vibration were made.

RESULTS

METHYL ETHYL KETONE

All results and descriptions of PBZ and GA air samples collected for MEK are shown in Table 8. Four short-term (15 minutes) PBZ air samples were collected for MEK during a ring cleaning operation in the Fittings Department. Exposures ranged from 149 to 421 ppm. Three of these PBZ samples exceeded the NIOSH STEL of 300 ppm for MEK. Overall, PBZ concentrations ranged from <10 ppm up to 421 ppm, TWA over the period sampled.

1,1,1-TRICHLOROETHANE

Results from all of the PBZ and GA air samples collected for 1,1,1-trichloroethane are shown in Table 9. Five short-term (15 to 19 minutes) PBZ air samples were collected for 1,1,1-trichloroethane during the *interior* cleaning of fuel cells (using this solvent) in the *Final Finish* Department. These exposures ranged from 293 to 878 ppm. Four of these PBZ samples exceeded the NIOSH STEL of 350 ppm for 1,1,1-trichloroethane. Overall, the PBZ concentrations ranged from <10 ppm up to 878 ppm, TWA over the period sampled.

ETHANOL

Three general area air samples were collected in the nylon spray operation to assess employee exposures to ethanol. As shown in Table 10, ethanol concentrations ranged from 26 to 160 ppm, expressed as TWAs over the period sampled. The NIOSH REL, OSHA PEL, and ACGIH TLV for ethanol is 1000 ppm for an 8- to 10-hour TWA exposure.

Health Hazard Evaluation Report No. 90-246

HEAT STRESS

The WBGT data collected on August 20, 1991, throughout the first (day) shift is presented in Table 11. The $WBGT_{in}$ measurements ranged from 75.4 to 83°F, with the dry bulb air temperature as high as 100.7°F and the radiant (globe) temperature reaching 101.4°F. These two highest temperatures were measured in the *Fittings* Department at approximately 2:30 p.m. The dry bulb temperatures outside ranged from 86 to 96.8°F on the day of this survey, weather conditions considered by the employees and management as mild for mid-summer.

The aural temperatures measured on an employee in the *Final Finish* Department are presented in Table 12. The aural temperatures remained fairly consistent over the approximately 2.5 hour sampling period, ranging from 36.8 to 37.1°C. None of the aural temperatures measured as part of this evaluation exceeded the maximum core body temperature of 38°C as proposed by NIOSH and the ACGIH.

MEDICAL

The employee information provided by union representatives included names and a brief account of health concerns for 37 current and former employees. Seven of the approximately 600 current employees were interviewed. In addition to these seven, 19 other current employees (not on the list initially provided by the union) were interviewed during our visit. Therefore, work-related health concern information was available for a total of 56 current or former employees. Figure 2 summarizes the health concerns of the 26 employees who were interviewed during this evaluation. All symptoms except for skin rash occurred in fewer than 25% of the individuals interviewed, and several (headaches, nervousness, and dizziness) were not suggestive of a particular medical diagnosis.

From a review of the OSHA 200 injury and illness records and interviews with employees, the most commonly reported health concern was skin rash. Workers attributed their skin rashes to contact with fiberglass, rubber cements, "shiny" gum on some of the rubber, solvent #6079, and MEK. Other suspected causes of the rashes included warm temperatures in the work area and buffing of cells (without skin protection). Several employees reported that protective clothing or gloves were not worn because they interfered with the ability to do the work and/or substances seeped through gloves.

Other substances that were specifically mentioned by those interviewed as contributing to a health concern included: solvent #12-400, which was reported to cause dizziness, nausea, bloating, and eye irritation; 1,1,1-trichloroethane, which reportedly causes dizziness and nausea; rubber "cements," the smell of which reportedly makes some persons dizzy and causes respiratory irritation; solvent #LQ-389, which

Health Hazard Evaluation Report No. 90-246

reportedly causes eye irritation; solvent #7172, which reportedly causes nausea and nervousness; and solvent #6079, which reportedly causes shortness of breath and nervousness.

Two women were reported to have been diagnosed with breast cancer. There was one reported case of cancer of the colon. Colon and breast cancer are common in the general population and the few cases at Amfuel do not provide reason to suspect a work-related cause for the occurrence of these cancers.

There were eight reports of other cancers, but no specific details were available regarding these. The individuals with these reported cancers were either no longer working at Amfuel or did not volunteer to be interviewed. Since the data for these other reported cancers were incomplete, it is difficult to determine whether there was any basis to suspect that they were associated with exposures at work.

ERGONOMICS

The following ergonomic findings are based on observations and analysis of videotape made during a site visit to two Amfuel facilities on August 19-22, 1991. Eight operations were included in this evaluation. Information such as the number of workers employed in each job, the types of tools used, the work station dimensions, and the force requirements of certain tasks was collected. A job analysis was performed to assess the repetitiveness of each task and to document instances of awkward hand, wrist, arm and trunk postures. Manual force requirements were estimated, and exposures to hand/arm vibration were also noted, although no direct measurements of hand-transmitted vibration were made.

FITTINGS DEPARTMENT

Four separate processes are performed in the fittings department.

1. **Rubber Cutting Area**

    Workers in the rubber panel cutting area process approximately 300 pieces (in batches of 5 or 10) per day. The steps in the operation are as follows:

    (a) One worker unrolls rubber sheets from a spindle and cuts a length from the roll using a pair of standard scissors;

    (b) The rubber sheet is placed on a die inside the press;

    (c) The operator activates the press by pushing two palm-buttons;

Health Hazard Evaluation Report No. 90-246

(d) After the press operation is completed, the operator removes the rubber from the die, trims the edges with a pair of scissors, and places it in a stack. The operator repeats steps 1-3 until each piece in the batch has been processed;

(e) The operator removes the die from the press and uses a small hand tool to remove rubber pieces from the die holes;

(f) The operator places the die on a storage shelf and retrieves a new die to be installed in the press for the next batch.

Ergonomic concerns:

(a) Rubber materials vary significantly in thickness and elasticity. Scissors are used almost exclusively for cutting rubber sheets into smaller panels. Depending on the thickness and elasticity of the material, cutting rubber sheets with a pair of scissors can require significant manual force exertion. The metal scissor handles also concentrate stress on the operator's thumb and fingers.

(b) Cutting large sheets of rubber into smaller panels frequently requires a long reach. Similarly, the reach to place and position sheets on the die is approximately 23 inches. In some cases (cutting rubber into thin strips is one example), operators reach over rollers to put materials on the die. Forward reaches of more than 20 inches should be avoided to limit static effort.

(c) On one occasion, after the press operation was completed, the operator was observed holding the die over a barrel and removing rubber scrap from die holes with a punch tool. While this operation occurs infrequently, it is potentially fatiguing and inefficient. Investigators noted that it is not unusual for workers to work up to twenty hours overtime each week. Improving efficiency should reduce overtime work, resulting in a savings to the company and greater recovery time between shifts for the employees.

2. **Build-up Area**

Workers in the build-up area cement layers of rubber material to each other and to metal fittings. Complex fittings are processed one-at-a-time; however simple pieces can be processed in batches of five to ten at once. Depending on the complexity of the fitting, cycle time/batch can vary from half-an-hour to a day or more.

Health Hazard Evaluation Report No. 90-246

Ergonomic concerns:

(a) Build-up operations require frequent finger pressing and pinching, application of force with the palm of the hand, and use of small hand tools which concentrate stress on the soft tissues of the hand. Scissors and small roller-equipped hand tools, known as "stitchers" are frequently used in the fittings department to remove air pockets caught between layers of rubber and metal. Workers roll the stitcher over the surface of the rubber while applying downward forces. Use of the stitcher results in ulnar and radial wrist deviation, and repetitive elbow and shoulder flexion/extension. Frequent and prolonged application of manual force is strongly linked to CTD development.

(b) The height of the work station and the lack of movable fixtures contribute to awkward shoulder and wrist postures. The height of the table top is 35.5 inches; however, fixtures located on top of the work station add four to five inches to the working height. Because much of the work requires large downward applications of force (see #1 above), operators were frequently observed working with the shoulders abducted and wrists in ulnar or radial deviation. Work fixtures were fashioned from wooden blocks (non-movable); therefore, trimming rubber from fittings with scissors resulted in twisting of the wrists and extreme ulnar wrist deviation with finger flexion.

(c) Workers stand throughout their work shift. Prolonged standing allows the blood to pool in the legs and feet and places a static load on muscles in the legs, back and trunk.

(d) The visual demands in the build-up area are quite high; operators are responsible for detecting and removing air pockets which can be very small and difficult to detect.

3. **Curing Area**

Fittings are placed in metal molds and cured in oven-presses for periods of 20-45 minutes. Two large oven-presses and as many as 12 small ovens are operated continuously in the curing area.

Ergonomic concerns:

Material handling activities are the primary source of ergonomic hazards in the curing department. Workers push and pull molds weighing from 30 to 100 pounds between ovens, platforms, carts and storage racks. Moving large

Health Hazard Evaluation Report No. 90-246

molds can require more than 50 pounds of force exertion with the back and upper extremities. Operators are also required to make excessive reaches across platforms and carts to pull molds from shelves or ovens. Shelf height varies from approximately 12 inches above the floor to 60 inches above the floor, causing the worker to execute pushes or pulls while bending or reaching above shoulder height. A chain-operated hoist is provided to help the operator lift covers from the tops of molds; however this device is somewhat cumbersome to operate.

4. **Finishing Area**

   The final process performed in the fittings area is finishing. Workers remove excess rubber from metal fittings using vibrating drills, buffing wheels, air guns, razor knives and grinders. Most work is performed at hooded work stations, designed to collect dust generated during grinding processes.

   Ergonomic problems:

   (a) Work in the finishing area requires almost continuous use of hand tools, many of which are capable of transmitting significant levels of vibration to the hand. Few operators were observed wearing gloves during grinding or buffing tasks. The diameter of the grinding tool handle also appeared to be too large for female operators' hands. Handles which are larger than the user's hand diameter require more effort to grip and manipulate than handles which match, or are somewhat smaller than the user's hand size.

   (b) Many of the operations in the finishing area require frequent and prolonged exertion of pinch grip forces. Razor blades (with and without handles) are frequently used to trim away small pieces of rubber. Operators in the final finishing area were observed using cretex bars to clean metal surfaces. Operators grip the cretex bar between the first two fingers and the thumb; therefore, use of the cretex bar is associated with forceful, sustained finger-pinching.

   (c) A loud, high-pitched noise emanating from the grinders was presented continuously to operators in the finishing area during the site visit. No workers in the finishing area were observed wearing hearing protection.

   (d) In one area of the finishing department, workers use large mounted buffing wheels to roughen the edges and surfaces of large rubber fittings (to improve bonding between the fitting and fuel cell). Although some fittings can be quite large and bulky, workers are required to hold and manipulate the fittings during this process, which often requires several minutes. Prolonged holding and application of pinch grip forces imposes a static load on the fingers, shoulders, and forearms. Static loading can result in soreness, loss of grip strength, and fatigue.

(e) All work in the finishing area is done while standing. Mats provided to cushion the work surface appeared thin and very worn.

RUBBER CUTTING DEPARTMENT

Although materials for fittings are cut in the *Fittings* department, panels for use in the *Innerliner* and *Outerply* Departments are cut in a separate, much larger area. Operations in the two areas are similar. Sheets of material are unrolled from a spindle to a predetermined length. Two operators, working from opposite sides of the cutting table, use standard (manual) scissors to cut the rubber sheet from the roll. Sheets are then stacked on a table. After a specified number of sheets have been cut, workers place a chalked template over the stack and rub chalk markings onto the top sheet. The operators use the chalk markings as a guide for cutting the large rubber sheets into smaller panels with power shears.

Two ergonomic concerns were noted in the *Rubber Cutting* department. First, cutting rubber sheets with a pair of manual scissors can require significant manual force exertion. The metal scissor handles concentrate stress on the operator's thumb and fingers. Although manual scissors are used less frequently in rubber cutting than in the *Fittings* or *Innerliner* departments, prolonged use could result in soreness or fatigue. Second, the work table used in the *Rubber Cutting* department is approximately five feet wide. Although two operators work from opposite sides of the table, neither can reach the middle without significant trunk flexion and an extended reach. Because cutting activities frequently require reaches to the middle of the table, operators spend much time leaning forward across the table with shoulders and arms extended.

INNERLINER AND OUTERPLY DEPARTMENTS

Fuel cells are largely constructed in the *Innerliner* and *Outerply* Departments, and these departments employ the largest number of workers. Operations in both areas are similar. In the *Innerliner* Department, the first layer of rubber is glued to a cardboard form (mandrel) in the shape of the fuel cell. Fittings are also applied to the fuel cell at this time. In the *Outerply* area, additional layer(s) of rubber are applied to the fuel cell. At the end of the process, the entire cell is placed in an autoclave to vulcanize (cure) the rubber and form a one piece unit.

Health Hazard Evaluation Report No. 90-246

Ergonomic concerns:

(a) Operations in the *Innerliner* and *Outerply* Departments require constant use of the hands. Applying rubber panels to the fuel cells requires finger pressing and pinching, application of force with the palm of the hand, and use of small hand tools which concentrate stress on the soft tissues of the hand. Manual scissors are used almost exclusively for cutting rubber sheets into smaller panels for attachment to the fuel cells. Depending on the thickness and elasticity of the material, cutting rubber sheets with a pair of scissors can require significant manual force exertion. The metal scissor handles also concentrate stress on the operator's thumb and fingers. The scissors used in the inner and outer ply departments appeared to be somewhat smaller than those used in cutting areas (almost like school scissors), meaning that more movements were required to make the same lengths of cuts. The rubber material often becomes very difficult to manipulate due to its size and tendency to wrinkle and stick together. Therefore, pulling and stretching the fabric with the hands and fingers is commonly observed. Frequent application of manual force is strongly linked to CTD development.

(b) Fuel cells vary tremendously in size and shape. Although some cells are small enough to rest on top of a table, others are 56 inches tall. Although some fuels cells are mounted on mandrels (which allow the operator to rotate the cell), the largest fuel cells are placed on sawhorse structures. Rotation requires the help of the maintenance crew. Working height in this configuration ranges from 13 to 69 inches. Significant bending and overhead reaching is required to completely cover the entire cell with panels. In almost all cases, the work surface is positioned vertically (perpendicular to the floor). Flat surfaces are much easier to work on; however, many fuel cells have highly irregular and convoluted surfaces. The frequency of cutting, pulling and stretching activities is increased as the fuel cell surface becomes more irregular.

(c) All work is done while standing. No mats or cushions were provided for operators to stand on while working.

NYLON SPRAY DEPARTMENT

An operator, using a trigger-activated spray gun, coats fuel cells with cement and/or nylon in the spray area. No conspicuous ergonomic problems related to the spraying task were identified.

Health Hazard Evaluation Report No. 90-246

CLEANING/INSPECTION AND REPAIR DEPARTMENTS

During final inspection, workers suspend the fuels cell from ropes. Workers partly or completely crawl inside the fuel cell to inspect for leaks, and remove any residue remaining from the production process. Repairs are made by applying sealant to any leaks, or clamping heating blocks to areas of the cell where air bubbles are trapped.

ONION TANK ASSEMBLY

Onion tank (i.e., 3000 gallon portable water tank) assembly operations are performed at a separate facility in Monticello, Arkansas. The onion tank assembly process consists of *four* separate operations: (1) panel cutting, (2) assembly of side panels and tank collars, (3) assembly of bottom panels, and (4) final assembly. Onion tanks also undergo integrity testing before they are packed in crates for shipping. All operations are performed at adjacent work stations within the same building.

Ergonomic concerns:

(a) Stitching tools are used extensively throughout the assembly areas. The manual stitching tools are similar to those used in the Magnolia plant, except that the roller surface is beveled (i.e., slanted with respect to the roller axis). The change in the orientation of the roller causes the operator to extend and ulnar deviate the wrist during use. In addition, a power stitcher is sometimes used for stitching tasks. The power stitcher is equipped with a pistol grip and a trigger. Demonstration revealed that the power stitcher emits significant levels of vibration during operation. Because the tool is used with the pistol grip parallel to the work surface, the operator is forced to abduct the elbow and ulnar deviate the wrist during use. Finally, to control the speed of the motor, the operator must alternately depress and release the trigger on a continuing basis.

(b) In most cases, workers stand beside tables to perform work tasks. Workers in the final assembly area, however, were observed working on top of the assembly table on hands and knees. Prolonged kneeling on hard surfaces can result in knee trauma.

(c) Workers in the packing and shipping area use a Signotde combination strapping tool to band wooden boxes together. Three bands are placed around each box. Operation of the strapping tool requires repetitive force exertion with the shoulder and arm. Further, boxes rest on the floor during

Health Hazard Evaluation Report No. 90-246

strapping; workers must bend at the waist to position and fix the straps in place.

DISCUSSION AND CONCLUSIONS

SOLVENT EXPOSURES

The PBZ air sampling results show that short-term (15 minutes) peak exposures to MEK and 1,1,1-trichloroethane occurred in excess of NIOSH criteria and OSHA exposure limits. Many of these episodes were related to the uncontrolled handling of these solvents or rubber cements by the workers without the benefit of local exhaust ventilation or other engineering controls. For example, workers in the *Cement House* and *Face Coater* areas were observed manually transferring MEK from 55-gallon drums into smaller plastic containers. These containers of solvent would then be used to clean brushes or other equipment or to mix batches of rubber cement. NIOSH investigators also observed employees throughout the facility transferring MEK-containing rubber cements from 5 or 10 gallon containers into disposable cardboard cartons or other receptacles. These activities were performed almost exclusively without the benefit of local exhaust ventilation (LEV).

In some instances employees handling MEK-containing cements and cleaners did not wear personal protective equipment such as impermeable gloves and/or aprons. In other situations workers who were exposed to MEK during operations such as ring cleaning and brush cleaning wore natural latex gloves and cloth aprons, materials which are not impermeable to this solvent.

Brushes are used by the workers in several department to apply the various rubber cements and other solvents to the fuel cells. During this NIOSH evaluation workers collected these brushes (and other miscellaneous hand tools) near the end of the day shift, sorted them (to eliminate the brushes which could not be reused), and then manually cleaned each brush using MEK. A short-term (19 minutes) air sample collected on the "brush cleaner" on 8/20/91 had 117 ppm of MEK. While the employee performing this cleaning job wore an air-purifying organic vapor respirator, he also used neoprene rubber gloves. Neoprene is not recommended for protection from MEK. Glove materials such as Teflon® or butyl rubber offer superior resistance and impermeability to MEK.

RESPIRATORY PROTECTION

NIOSH investigators reviewed the written respirator program developed by Amfuel. While the main elements of any respiratory protection program were present (such as employee training, respirator selection, fit-testing) the overall program was judged inadequate based on the following findings:

Health Hazard Evaluation Report No. 90-246

- <u>Improper Respirator Selection</u>. Half-face piece organic vapor air purifying respirators were provided and worn by employees in the *Final Finish* Department for protection against 1,1,1-trichloroethane. Personal exposures to this solvent were measured in excess of the NIOSH STEL. Since NIOSH considers 1,1,1-trichloroethane to exhibit poor warning properties at concentrations below the REL, only supplied air or self-contained breathing apparatus (SCBAs) are recommended as suitable respiratory protection.

- <u>Respirator Maintenance</u>. Reportedly, most Amfuel employees changed the cartridges of their organic vapor respirators on a weekly schedule. However, because of the limited useful service time of organic vapor cartridges (or canisters), NIOSH recommends that they be replaced daily or after each use, or even more often if the wearer detects odor, taste, or irritation. Discarding the cartridge/canister is recommended at the end of the day, even if the wearer does not detect odor, taste, or irritation.

- <u>Improper Respirator Use</u>. NIOSH investigators observed several workers with beards wearing half-mask air purifying respirators. Facial hair can interfere with the facial seal of the respirator and prevent the wearer from obtaining a proper fit. Several employees were observed in the *Final Finish* Department cleaning the interior of small fuel cells by reaching inside the cell. During this cleaning process these employees would typically have their head and/or upper half of their body inside the cell. The cleaning solvent was 1,1,1-trichloroethane. Short-term (15-minute) exposures up to 878 ppm were measured on several of these workers. It was company policy that respirators were not required for these employees since they could remain outside the fuel cell with only their head (or upper body) inside the cell during the cleaning operation.

CONFINED SPACES

Using the definition found in the NIOSH criterion for working in confined spaces (*limited access; unfavorable natural ventilation; not intended for continuous worker occupancy*), the fuel cells which the employees must enter in the *Final Finish* Department are confined spaces.[19] Air sampling results from this NIOSH evaluation measured personal exposures which approached, but did not exceed, the NIOSH Immediately Dangerous to Life or Health (IDLH) level of 1000 ppm for 1,1,1-trichloroethane.[b] However, in the OSHA health and safety inspection which

---

[b] Half-mask organic vapor respirators were worn by the employees in the *Final Finish* Department when they were using 1,1,1-trichloroethane inside the fuel cells. The NIOSH respiratory protection guidelines do not recommend air-purifying respirators for protection from overexposures to 1,1,1-trichloroethane.

preceded this NIOSH evaluation, PBZ exposures in excess of 1000 ppm were measured among employees working inside the fuel cells in the Final Finish Department. Based on the following three reasons, the interior of the fuel cells should be classified as "**Class A**" confined spaces (See Table 6) whenever 1,1,1-trichloroethane is being used by the worker inside the cell. It is important to note that this confined space classification would apply regardless of the location or orientation of the fuel cell (for example, whether the cell in suspended from an elevated work table, sitting on the floor, etc.).

- PBZ exposures will vary depending on a number of factors, including the type of work task being performed, the work techniques utilized, the environmental conditions (temperature, humidity) during the sampling period, etc.

- Solvent exposures measured by NIOSH investigators in this evaluation ranged up to 878 ppm for a 15-minute STEL. However, air sampling data collected by OSHA compliance officers in 1990 measured airborne levels of 1,1,1-trichloroethane of up to 3536 ppm (STELs) during interior fuel cell cleaning in the *Final Finish* Department.

- Air-purifying organic vapor respirators, which do not offer adequate protection against 1,1,1-trichloroethane, were being worn by the employees when they were using this solvent for interior cleaning in the cells.

If employees are working inside the cells and no solvents (such as 1,1,1-trichloroethane) are being used, then the work area could be classified as a "**Class C**" confined space. Table 7 contains a check list of considerations for entry, working in, and exiting confined spaces.

ERGONOMICS

Some of the specific ergonomic recommendations include replacing manual cutting shears with powered shears; providing a fixture or tool to remove scrap rubber from the hole-cutting die; re-designing scissors with longer handles, shorter blades, and a self-opening mechanism to reduce the manual stress associated with prolonged and repetitive tool use; changing the height of the work stations to reduce the occurrence of awkward wrist postures and long reaches; providing stools, cushioned floor mats, or raised foot rests; providing additional lighting and (where needed) magnifying glasses to improve visibility; adding rollers to the bottom surface of platforms and racks to allow the operator to transfer molds between surfaces with less force exertion; reducing tool vibration; and modifying tool handles to eliminate conditions which require a pinch grip.

Health Hazard Evaluation Report No. 90-246

MEDICAL

Colon and breast cancers are common in the general population and the few cases at Amfuel did not suggest an occupational etiology. Since the data for the other reported cancers were incomplete, it is impossible to estimate the likelihood that they were associated with exposures at work. However, unless these cancers are in unusual sites or are all in the same site, it is unlikely that they would contribute evidence for being related to work.

The skin irritation reported by Amfuel employees is not unexpected. Both chemical and mechanical trauma to the skin has been reported among workers in similar rubber fabricating facilities.[20]

HEAT STRESS

Although not evident during the follow-up survey conducted in August 1991 during which WBGT measurements were obtained, it is possible that more severe heat stress conditions could exist during periods of warmer weather. Departments such as *Final Finish* (working inside the fuel cells) and *Fittings* (the use of heated presses to assemble the metal/rubber components) would likely be the most severely affected.

RECOMMENDATIONS

Since the recommendations from this evaluation cover a variety of areas, they have been grouped into categories.

ERGONOMICS

In all cases, engineering controls are the preferred method of reducing CTD risk. The goal of engineering controls is to make the job fit the person, not the person fit the job. Administrative (personnel-based) controls should be used only as a temporary measure to control CTD risk until engineering changes can be implemented. In addition, a medical management program for CTDs should be implemented. This program should provide mechanisms for identifying CTD cases at an early stage and providing treatment to employees before problems become more serious. Light duty assignments should be identified to allow workers continue work until CTD symptoms can be resolved.

The following ergonomic recommendations pertain to activities performed in the *Fittings, Innerliner, Outerply, Final Inspection, Nylon Spray, and Onion Tank Assembly* areas.

Health Hazard Evaluation Report No. 90-246

1. *Rubber Cutting Area*

    (a) **Implement an alternative method for cutting the rubber sheets.** The manual scissors currently in use should be replaced with powered shears to facilitate cutting. An alternative is to modify the rack holding the rolls of material to include a track-mounted cutting wheel, similar to those used to cut lengths of material in fabric stores. Powered shears are currently being used to cut materials used in other parts of the plant.

    (b) **Modify the method in which the press job is performed to reduce the reach distance to the back edge.** For operations where only one pass of the roller is required to complete the cycle, the worker should be instructed to return the roller to the back of the press before unloading finished pieces. This practice would eliminate reaches over the roller, and potential accidents involving the roller.

    (c) **Provide a fixture or tool to remove scrap rubber from the hole-cutting die.** A large ring- or U-shaped tool with small rod-like extensions around the perimeter could be used to remove scrap rubber from the die. The spacing of the rods would match the holes on the die. An alternative would be a fixture that the die could be placed on which would punch the scrap out of each hole simultaneously.

2. *Rubber Build-up Area*

    (a) **Alternative hand tools are needed to reduce the manual stress associated with prolonged and repetitive tool use.** Specifically, scissors and stitchers should be modified to reduce awkward hand/wrist postures and stress concentrations on soft tissue areas. Scissors should be provided with longer handles (to distribute the stress over several fingers and a larger area of the palm), a shorter blade (to reduce elbow and shoulder abduction), and a self-opening mechanism (to reduce the force required to operate the scissors). Stitchers should have padded handles to reduce hand stress during continuous gripping. The handle should also be angled with respect to the roller to reduce wrist deviation.

    (b) **For tasks requiring large downward forces, the working height of the hands should be 36 inches.**[8] Therefore, the height of the work station should be lowered when fixtures are used to accommodate for the extra height. Fixtures which allow free movement of the part are needed to reduce the occurrence of awkward wrist postures and long reaches.

(c) **Provide stools, cushioned floor mats, or raised foot rests**. These items would allow workers to rest leg and back muscles during prolonged periods of standing.

(d) **Additional lighting and mounted, swivel-type, magnifying glasses (such as those used in electronics assembly) are recommended to improve visibility in the build-up area**. Improving visibility would not only reduce neck flexion and eye strain, but should also improve product quality.

3. *Curing Area*

Recommendations include adding rollers to the bottom surface of platforms and racks. Rollers would allow the operator to transfer molds between surfaces with less force exertion. Stop bars, guides, or barriers would be needed along the edges of platforms and racks to make sure molds wouldn't roll when unattended. Use of the top and bottom storage shelves should be eliminated to eliminate excessive low or high reaches. Handles should be extended from the side of molds to reduce the reach required to remove molds from racks or ovens to the transportation platform.

4. *Finishing Area*

(a) **Interventions to reduce the tool vibration are needed**. At a minimum, operators should be provided with padded gloves, or the tool handle should be covered with a vibration-absorbing material (e.g., sorbothane). A regular tool maintenance program is also keep tool vibration levels at a minimum. Although not a permanent solution, rotating workers to other jobs which do not require vibrating tool use should be considered to limit exposure. Also, tools with smaller handle diameters are needed to reduce manual effort requirements. A handle diameter of 1.5 inches should better accommodate the majority of workers in the fittings area.

(b) **Modifications to tools handles are needed to eliminate conditions which require a pinch grip**. Razor blades should be mounted on handles to reduce pinch grip force and the risk of accidental cuts. It is recommended that an alternative to the cretex bar be used for cleaning metal and rubber surfaces. Specifically, a larger sanding block would to allow operators to hold the device with a power grip instead of a pinch grip.

Health Hazard Evaluation Report No. 90-246

(c) **Testing is recommended to determine if noise exposure is excessive.** If so, hearing protection should be made available to employees, and its use should be encouraged.

(d) **Replace pedestal-mounted grinders with hand-held models.** Buffing and grinding operations can be performed more easily using a hand-held grinder against a mounted work piece. A fixture should be provided to hold fittings while workers perform grinding operations with hand-held tools. Precautions (as discussed in (1) above) should be taken to limit vibration transmission by these tools.

(e) **Provide stools or sit/stand chairs to reduce static loading of leg, back and shoulder muscles.** At a minimum, cushioned mats should be replaced regularly, and a footrest should be provided at the front of the work station.

5. *Rubber Cutting Department*

    Two recommendations to eliminate hazard in the *Rubber Cutting* department are provided. First, it is not clear why workers do not use the power shears to cut the rubber into sheets. The task could be performed just as quickly and efficiently with the power shears, and with less manual stress. A button-operated "guillotine" device could also be implemented; this option would probably be more expensive, but might allow the task to be performed by one operator instead of two. Second, an approach similar to that used in the fittings department could be used to cut the rubber sheets into smaller panels. Specifically, a die (albeit a large one) would be placed on the table before cutting operations were initiated. Sheets of rubber would be placed over the die; a large roller would then press the rubber into the die, cutting the sheets into smaller pieces. Since this operation would be largely automated (except for positioning the sheets correctly) it could, again, be performed by a single operator. Automating the process would also ensure more uniformity in sheet size. Further, the awkward reaches associated with the manual cutting process would be eliminated.

6. *Innerliner and Outerply Departments*

    (a) **Minimize stress associated with tool use and work station configuration.** It is unclear, however, that an alternative exists for a less hand-intensive method of applying rubber to the fuels cells. As a beginning, the stitchers should be modified to reduce awkward hand/wrist postures and stress concentrations on soft tissue areas. Stitchers should have padded handles (to reduce hand stress during

Health Hazard Evaluation Report No. 90-246

continuous gripping), and the handle should be angled with respect to the roller to reduce wrist deviation. Scissors and stitchers should be modified to reduce awkward hand/wrist postures and stress concentrations on soft tissue areas. Scissors should be provided with longer handles (to distribute the stress over several fingers and a larger area of the palm), a longer blade (to reduce the number of cuts needed to separate panels), and a self-opening mechanism (to reduce the force required to operate the scissors).

(b) **Provide anti-fatigue mats in areas where workers stand to work on fuel cells.** Low stools should also be provided for operators working on lower portions of the fuel cell. A sit/stand chair can be used when workers spend longer periods of time working on one area of the fuel cell.

7. *Nylon Spray Department*

Movement of mandrels through the spray area did appear to present some potential material handling problems however. The wheels on the mandrels are fairly small (diameter unknown); the floor covering in the spray area is cracked and pitted in some locations. Under best case conditions (wheels straight, good floor), a peak force of 25-35 pounds is required to initiate movement of the mandrels from a static position. This force increases significantly if the floor is uneven, or the rollers are not pointed in the right direction. Because excessive push/pull forces can result in back strain or overexertion; a regular floor maintenance program should be initiated, and larger wheels should be installed on mandrels.

8. *Cleaning/Inspection and Repair Departments*

Although inspection and repair processes are somewhat awkward (e.g., climbing in and out of the fuel cells), they are not highly forceful or repetitive. Therefore, the associated hazards are probably relatively minor. Recommendations include providing inspectors with magnifying glasses to improve their ability to detect defects.

9. *Onion Tank Assembly*

(a) **While powered tools are generally less stressful to use than manual tools, it is doubtful that the power stitcher represents an improvement in stitcher design.** Because the power stitcher produces significant levels of vibration, its use is not recommended. However, if the power stitcher must be used, modifications in its design are recommended. First, the pistol grip should be replaced with a straight handle, to allow the user to maintain a neutral wrist and shoulder position. The handle should also be covered with a material to absorb vibration and cushion the grip. Second, the finger-activated trigger should be replaced with a strip trigger, which distributes force over a larger surface area.

(b) **Floor mats and railing along the bottoms of work tables are needed to relieve foot and leg stress in workers who stand continuously.** Workers who perform final assembly tasks should be provided with knee and elbow pads, to reduce the potential for trauma during kneeling. A short ladder or stepping stool is needed to help workers climb up to the table top.

(c) **An automatic box strapper should be used to band wooden boxes together.** Boxes would be placed on a conveyor belt which would pass through the box strapper; the strapping machine would automatically position the straps and tighten the bands to the correct tension. This machine would not only eliminate the bending and repetitive force exertion associated with the manual strapping device, but would also make the process more efficient.

WORK PRACTICES

1. Gloves and protective clothing should be selected based on their permeation and degradation resistance to the solvents being used by the worker. NIOSH investigators observed employees using gloves made of natural latex or neoprene when handling solvent such as MEK or 1,1,1-trichloroethane. These glove materials are not recommended for protection from these solvents. Examples of materials which offer superior protection include Teflon® or butyl rubber (for MEK exposures) and Viton®, Teflon®, or polyvinyl alcohol (for 1,1,1-trichloroethane exposures). While these glove materials offer better permeation and degradation resistance than natural latex, a glove's resistance to cuts, snags abrasion, punctures, or tears must also be considered. Another factor is an adequate sleeve (or cuff) length to protect the forearm from solvent exposure. Actual workplace conditions may require a combination of performance capabilities.

2. Employee exposures to MEK during brush cleaning could be reduced by changing or modifying the cleaning procedure. For example, a greater use of disposable (single-use) brushes by Amfuel should reduce the number of brushes which must be manually cleaned. The use of LEV to control solvent

Health Hazard Evaluation Report No. 90-246

emissions during the brush cleaning operation would also reduce employee exposures.

3. The cleaning and inspection procedures followed by employees in the *Final Finish* Department should be examined with the intent of reducing workers' solvent exposures. For example, the use 1,1,1-trichloroethane during the *initial* interior cleaning and inspection of a fuel cell could be eliminated. The cell could be air-leak tested prior to the *second (final)* interior cleaning with this solvent.

CONFINED SPACES

A Confined Space program should be developed and implemented consistent with the guidelines contained in DHEW (NIOSH) Publication No. 80-106, "Working in Confined Spaces." This program would contain the minimum program elements for a **Class A** confined space as listed in Table 6. The interior of fuel cells would be considered confined spaces since these areas have limited access, unfavorable natural ventilation, and not intended for continuous worker occupancy. The interior of these fuel cells should be considered confined spaces regardless of the cell's spatial orientation (for example, sitting on the floor or suspended on a work table) or whether the worker is required to completely enter the cell to perform the work.

RESPIRATORY PROTECTION

1. Amfuel's written respiratory protection program (dated 6/15/91) contains most of the major program elements required by OSHA General Industry Standard 29 CFR Part 1910.134 and the NIOSH respirator guidelines. However, the program was not effective in several important areas, such as proper respirator selection and employee training. The written program should be revised to designate one individual with the responsibility for administering the respiratory protection program. Currently this responsibility is shared between the Personnel Department, the Production Supervisors, and the Environmental Coordinator. The written respirator program should also contain information on the following topics: (1) the departments/operations which require respiratory protection; (2) the correct respirator(s) required for each job/operation; (3) specifications that only NIOSH approved respiratory devices shall be used; and (4) the criteria used for the proper selection and use of respirators, including limitations. The Amfuel respirator program should also reference the requirements contained in the Confined Space program to assure that employees are adequately protected when working in these areas.

2. Based on NIOSH respirator selection criteria, an organic vapor/air purifying respirator is **not recommended** for protection against exposures to

Health Hazard Evaluation Report No. 90-246

1,1,1-trichloroethane. This is based on the uncertainty regarding the odor threshold for this compound. Either a supplied air (SA) respirator or a self-contained breathing apparatus (SCBA) is recommended for exposures to 1,1,1-trichloroethane of up to 1000 ppm.[c] At levels above 1000 ppm, the NIOSH respirator selection criteria require (as a minimum) full-face piece SCBAs operating in a pressure-demand mode and equipped with emergency escape provisions.

HEAT STRESS

1. A written heat stress control policy and program which addresses the topics listed in the NIOSH document *Criteria for a Recommended Standard: Occupational Exposure to Hot Environments*, should be developed. Development of a heat alert program and medical surveillance should be incorporated into the heat stress program.

2. Additional monitoring should be conducted to determine the extent that warmer (and/or more humid) days will impact on heat stress conditions in these departments. These and other heat stress program elements are listed in Appendix A.

3. The use of portable spot coolers should reduce the heat stress to *Final Finish* employees while they work inside the fuel cells. Several models offer portability (the cooling unit can be mounted on casters), versatility (the cold air can be directed to the hot spots via flexible tubing), and ease of operation (some models operate from a standard 115 volt power supply and require only an occasional filter change). These cooling devices would also provide the necessary uncontaminated dilution air to the interior of the fuel cells.

MEDICAL

1. Physicians and other health care personnel who may provide medical care to Amfuel employees should be provided with pertinent information which would help to characterize the exposure potential for that worker to hazardous materials. For example, industrial hygiene air sampling data, personal protective equipment (if any) worn by the worker, and a listing of hazardous materials used by the employee while at work, would provide useful information to the health care provider in selecting the appropriate medical

---

[c] Based on NIOSH recommendations, an immediately dangerous to life or health (IDLH) atmosphere exists upon exposures to 1,1,1-trichloroethane of 1000 ppm.

Health Hazard Evaluation Report No. 90-246

surveillance for that worker. If respiratory protective equipment is determined to be necessary, medical evaluations should be conducted to determine the worker's physical fitness for using this equipment. In addition, complete medical, chemical exposure, and occupational history information should be maintained for each worker.

2. Amfuel should provide a worker education program designed to inform the worker about the potential health risks from exposure to hazardous substances, the proper use of personal protective equipment or clothing, and proper work practice procedures. This should involve more than simply handing out literature for the employees to read. Health care personnel and/or others knowledgeable about these issues should discuss each of these topics with the employees, allowing adequate time for questions.

3. Although initially unrelated to this health hazard evaluation request, exposure to environmental tobacco smoke (ETS) is an important public health problem. Reports from the Surgeon General, the National Research Council and EPA have concluded that exposure to ETS may be associated with a wide range of health (e.g., lung cancer) and comfort (e.g., eye, nose, and throat irritation and odor) effects.[21-25] NIOSH has determined that ETS may be related to an increased risk of lung cancer and possibly heart disease in occupationally exposed workers who do not smoke themselves.[26]

A smoking cessation program may be necessary to assist those employees who are current smokers. If smoking is permitted within the facility, it should be restricted to designated smoking areas.[26] These areas should be provided with a *dedicated exhaust system* (room air directly exhausting to the outside), an arrangement which eliminates the possibility of re-entrainment and recirculation of any secondary cigarette smoke. In addition, *the smoking area should be under negative pressure relative to surrounding occupied areas*. The ventilation system supplying the smoking lounge should be capable of providing at least 60 cfm of outdoor air per person. This air can also be obtained from the surrounding spaces (transfer air).

VENTILATION

The following recommendations are based on job-specific observations which were made during the walk-through ventilation evaluation performed in 1991. Additionally, Appendix B contains examples of ventilation designs applicable to a variety of industrial operations, including spray painting, grinding, and buffing. These ventilation designs were obtained from *Industrial Ventilation: A Manual of Recommended Practice* (19th Edition), a document published by the ACGIH.

Health Hazard Evaluation Report No. 90-246

1. *Ring Wash and Pot Wash.* The effectiveness of the current slot ventilation design can be increased by enclosing the sides.

2. *Ring Reclaim.* This operation is infrequently performed and is located outdoors. Considering these factors, the current requirement to continuously operate the three propeller-type fans mounted adjacent to the ring reclaim tank to is probably unnecessary. The ring reclaim operator, however, would receive a greater reduction in his/her exposures by using a remotely operated electric or pneumatic hoist to increase the distance between the worker and the reclaim tank. In addition, the tank temperature should be lowered prior to parts removal to decrease tank emissions.

3. *Fittings Department.* The elimination of the excessive lengths of flexible exhaust duct will improve the LEV effectiveness at the pedestal grinders and buffers. Sharp angled bends in the duct should be avoided. Additionally, since flexible duct has greater air resistance, it should not be used in place of smoother, solid-wall duct. A LEV system should be designed for the final finishers to capture the fine metal dust generated during this operation (there was no LEV provided for this procedure during this NIOSH survey). Transport velocity in ducts which carry particulates should be > 3500 feet per minute (fpm) to avoid settling.

4. *Spray Booth (Neoprene).* The existing panel fan should be replaced with a new blower and duct to increase the exhaust ventilation capability of the spray booth. The new duct should extend above roof height to prevent recirculation back into the plant.

5. *Spray Booth (Cement Room).* The fire door, located adjacent to this booth, should be kept closed. Local exhaust ventilation should be considered to control spurious emissions from the cement and solvent container stored adjacent to the booth. To avoid unnecessary solvent exposures, employees should be encouraged to keep the workpiece between them and the exhaust during spraying.

6. *Hydrohone (Fittings Department).* The air intake holes (plugged with fiberglass insulation during the NIOSH follow-up evaluation) must be kept open to permit air to enter the abrasive blasting chamber. The entire bag-house filter assembly should be relocated outside the building to minimize dust generation inside the building during cleaning. This filter assembly disconnects easily from the abrasive blasting cabinet. In addition, the exhaust system should be kept running whenever the cabinet is opened.

Health Hazard Evaluation Report No. 90-246

7. *Vapor Degreaser (Fittings Department)*. The exhaust ducts from the degreaser should be extended above the roof line of the building to prevent recirculation back into the building. The water temperature should be monitored to assure that chilled water is being provided to the degreaser coils. An automatic temperature monitoring system connected to an indicator alarm is recommended for the degreaser.

8. *Face Coater*. The canopy hood at the take-off end of the face coating line should be extended to enclose the face coating operation as much as possible. Slot ventilation should be added along both sides of the cement dip tray and this area should be enclosed as much as possible.

9. *Cutting Room (Innerliner Department)*. Repair bent slot ventilation on either side of the cutting table. The height of these slots should be extended above the height of the material on the table. Following this repair and redesign, the slot ventilation system for this operation should be balanced.

10. *Tab Assembly (Innerliner Department)*. The drive belt on the overhead paddle fan should be tightened. This fan should be repositioned to direct its air movement slightly downward and toward the tab assembly operation.

11. *Cement Room*. Slot exhaust hoods should be installed which are large enough to accommodate "brush washing," the weekly cleaning of the cement pots, and the small batch mixing of the Uniroyal® cements.

12. *Miscellaneous*. The use of fresh air showers in areas where solvents are used should be investigated. These devices provide localized clean air at low velocity around the employee. The air's low velocity minimizes the mixing of the clean air with contaminated air.

REFERENCES

1. CDC [1992]. Compendium of NIOSH recommendations for occupational safety and health standards. Atlanta, GA: U.S. Department of Health and Human Services, Public Health Service, Centers for Disease Control, National Institute for Occupational Safety and Health.

2. ACGIH [1992]. Threshold limit values and biological exposure indices for 1992-93. Cincinnati, OH: American Conference of Governmental Industrial Hygienists.

3. Code of Federal Regulations [1989]. OSHA Table Z-1. 29 CFR 1910.1000. Washington, DC: U.S. Government Printing Office, Federal Register.

Health Hazard Evaluation Report No. 90-246

4. U.S. DOL [1991]. BLS reports on survey of occupational injuries and illnesses in 1991. Washington D.C.: U.S. Department of Labor, Bureau of Labor Statistics, U.S. DOL News.

5. Armstrong TJ [1986]. Ergonomics and cumulative trauma disorders. Hand Clinics 2:553-565.

6. Putz-Anderson, V (ed) [1988]. Cumulative trauma disorders: a manual for musculoskeletal diseases of the upper limbs. New York, NY: Taylor & Francis.

7. Silverstein BA, Fine LJ, Armstrong TJ [1987]. Hand-wrist cumulative trauma disorders in industry. British Journal of Industrial Medicine 43:779-784.

8. Armstrong TJ, Foulke J, Joseph B, Goldstein S [1982]. Investigation of cumulative trauma disorders in a poultry processing plant. American Industrial Hygiene Association Journal 43:103-116.

9. Hales T, Habes D, Fine L, Hornung R, Boiano J [1989]. John Morrell & Co., Sioux Falls, SD; HETA Report 88-180-1958. Cincinnati, OH: U.S. Department of Health and Human Services, Public Health Service, Centers for Disease Control, National Institute for Occupational Safety and Health, Division of Surveillance Hazard Evaluation and Field Studies.

10. Habes DJ, Putz-Anderson V [1985]. The NIOSH program for evaluating hazards in the workplace. Journal of Safety Research 16:49-60.

11. Yaglou C, Minard D [1957]. Control of heat casualties at military training centers. Arch Indust Health 16:302-316.

12. Belding H, Hatch T [1955]. Index for evaluating heat stress in terms of resulting physiological strain. Heat Pip Air Condit 27:129.

13. Houghton F, Yaglou C [1923]. Determining lines of equal comfort. J Am Soc Heat and Vent Engrs 29:165-176.

14. NIOSH [1986]. Criteria for a recommended standard: occupational exposure to hot environments, revised criteria. Cincinnati, OH: U.S. Department of Health and Human Services, Public Health Service, Centers for Disease Control, National Institute for Occupational Safety and Health, DHHS (NIOSH) Publication No. 86-113.

Health Hazard Evaluation Report No. 90-246

15. ACGIH [1990]. Threshold limit values and biological exposure indices for 1990-1991. Cincinnati, OH: American Conference of Governmental Industrial Hygienists.

16. NIOSH [1972]. Criteria for a recommended standard: occupational exposure to hot environments. Cincinnati, OH: U.S. Department of Health, Education and Welfare, Health Services and Mental Health Administration, National Institute for Occupational Safety and Health, DHEW (NIOSH) Publication No. 72-10269.

17. Kenney WL [1985]. A review of comparative responses of men and women to heat stress. Environ Res 37:1-11.

18. NIOSH [1977]. Assessment of deep body temperatures of women in hot jobs. Cincinnati, OH: U.S. Department of Health, Education, and Welfare, Public Health Service, Health Services and Mental Health Administration, National Institute for Occupational Safety and Health, DHEW (NIOSH) Publication No. 77-215.

19. NIOSH [1979]. Working in confined spaces. Cincinnati, OH: U.S. Department of Health, Education, and Welfare, Public Health Service, Health Services and Mental Health Administration, National Institute for Occupational Safety and Health, DHEW (NIOSH) Publication No. 80-106.

20. Klemme JC [1985]. Chemical and mechanical trauma to the skin in a rubber fabrication facility. American Journal of Industrial Medicine 8:355-362.

21. HEW [1979]. Smoking and health: a report of the Surgeon General. Office on Smoking and Health. Washington, D.C.: U.S. Department of Health, Education, and Welfare, U.S. Government Printing Office.

22. HHS [1982]. The health consequences of smoking -- cancer: a report of the Surgeon General. Office on Smoking and Health. Washington, D.C.: U.S. Department of Health and Human Services, U.S. Government Printing Office.

23. HHS [1984]. The health consequences of smoking -- chronic obstructive lung disease: a report of the Surgeon General. Office on Smoking and Health. Washington, D.C.: U.S. Department of Health and Human Services, U.S. Government Printing Office.

24. HHS [1983]. The health consequences of smoking -- cardiovascular disease: a report of the Surgeon General. Office on Smoking and Health.

Health Hazard Evaluation Report No. 90-246

Washington, D.C.: U.S. Department of Health and Human Services, U.S. Government Printing Office.

25. HHS [1986]. The health consequences of involuntary smoking: a report of the Surgeon General. Office on Smoking and Health. Washington, D.C.: U.S. Department of Health and Human Services, U.S. Government Printing Office.

26. CDC [1991]. NIOSH current intelligence bulletin 54-environmental tobacco smoke in the workplace (lung cancer and other effects). Cincinnati, OH: U.S. Department of Health and Human Services, Public Health Service, Centers for Disease Control, National Institute for Occupational Safety and Health, DHHS (NIOSH) Publication No. 91-108.

AUTHORSHIP AND ACKNOWLEDGEMENTS

Report Prepared by:
Gregory A. Burr, C.I.H.
Supervisory Industrial Hygienist
Industrial Hygiene Section
Hazard Evaluations and Technical
 Assistance Branch
Division of Surveillance, Hazard
 Evaluations, and Field Studies

Katharyn A. Grant, Ph.D.
Industrial Engineer
Psychophysiology and
 Biomechanics Section
Applied Psychology and
 Ergonomics Branch
Division of Biomedical and
 Behavioral Science

A. Yvonne Boudreau, M.D.
Medical Officer
Medical Section
Hazard Evaluations and Technical
 Assistance Branch
Division of Surveillance, Hazard
 Evaluations, and Field Studies

Matthew K. Klein, P.E.
Research Mechanical Engineer

Health Hazard Evaluation Report No. 90-246

Field Assistance by:
Industrial Hygiene Section

Steve Lenhart, C.I.H.
Senior Industrial Hygienist
Industrial Hygiene Section

John A. Decker, M.S.
Industrial Hygienist
Industrial Hygiene Section

Leslie Copeland, M.S.
Visiting Scientist
Medical Section
Hazard Evaluations and Technical
Assistance Branch
Division of Surveillance, Hazard
Evaluations, and Field Studies

Originating Office:
Hazard Evaluations and Technical
Assistance Branch
Division of Surveillance, Hazard
Evaluations, and Field Studies

DISTRIBUTION AND AVAILABILITY OF REPORT

Copies of this report may be freely reproduced and are not copyrighted. Single copies of this report will be available for a period of 90 days from the date of this report from the NIOSH Publications Office, 4676 Columbia Parkway, Cincinnati, Ohio 45226. To expedite your request, include a self-address mailing label along with your written request. After this time, copies may be purchased from the National Technical Information Service, 5285 Port Royal Road, Springfield, VA 22161. Information regarding the NTIS stock number may be obtained from the NIOSH Publications Office at the Cincinnati address.

Copies of this report have been sent to:

1. Amfuel, Inc., Magnolia, Arkansas
2. United Rubber Workers, International Union, Akron, Ohio
3. United Rubber Workers, Local 607, Waldo, Arkansas
4. OSHA, Region VI
5. NIOSH

Health Hazard Evaluation Report No. 90-246

For the purpose of informing affected employees, copies of this report shall be posted by the employer in a prominent place accessible to the employees for a period of 30 calendar days.

TABLE 2
Toxicity and Permissible Exposure Information
HETA 90-246
Amfuel, Inc.

SUBSTANCE	TOXICITY INFORMATION	EXPOSURE CRITERIA
Methyl chloroform (1,1,1-trichloroethane)	Clear, non-flammable liquid. Oral toxicity of this solvent is low. Although skin absorption can occur, it is not considered a significant exposure route. Methyl chloroform is an anesthetic and, like many organic solvents, will defat the skin, causing dryness, redness, and scaling of the exposed skin. This solvent is poorly metabolized once in the body and is excreted unchanged in the expired air. Deaths due to anesthesia and/or cardiac sensitization has been observed in industrial exposures involving poorly ventilated or confined areas. In some studies involving human exposures, anesthetic effects were observed at concentrations approaching 500 ppm. In a long-term study of workers exposed to methyl chloroform (at concentrations which in some situations exceeded 200 ppm) no adverse effects related to exposure were observed.	NIOSH REL = 350 ppm (ceiling limit) OSHA PEL = 350 ppm (TWA) 450 ppm (short-term exposure limit) ACGIH TLV = 350 ppm (TWA) 450 ppm (short-term exposure limit)
Methyl ethyl ketone	MEK, a colorless, flammable liquid with a low odor threshold is a widely used industrial solvent. The threshold for eye and nose irritation is estimated at approximately 200 ppm. Most people can smell MEK at a concentration of 10 ppm. With the exception of complaints about its objectionable odor (which resembles acetone), few serious health effects have been observed under typical industrial exposure conditions. In addition to being absorbed through the skin, prolonged skin contact can lead to dermatitis. In one study workers exposed to airborne MEK concentrations of 300 to 600 ppm (along with skin contact to MEK) complained of numbness in the upper extremities. MEK is eliminated either in the expired air (unchanged) or in the urine (metabolized).	NIOSH REL = 200 ppm (TWA) 300 ppm (short-term exposure limit) OSHA PEL = 200 ppm ACGIH TLV = 200 ppm (TWA) 300 ppm (short-term exposure limit)
Ethanol	Also called ethyl alcohol, this solvent is flammable, colorless, and possesses a distinct odor. Under typical industrial exposure conditions, the acute toxicity of ethanol is low. Effects resulting from over-exposure to ethanol may include incoordination and drowsiness. Eye and skin irritation may result following contact with the liquid. In its vapor form ethanol is irritating to the eyes and upper respiratory tract at concentrations well below the established exposure criteria.	NIOSH REL = 1000 ppm (TWA) OSHA PEL = 1000 ppm (TWA) ACGIH TLV = 1000 ppm (TWA)

Abbreviations:

REL = Recommended Exposure Limit   TLV = Threshold Limit Value   ppm = part per million
PEL = Permissible Exposure Limit   STEL = Short-term exposure limit   TWA = Time-weighted average

References:

1. ACGIH [1986]. Documentation of the threshold limit values and biological exposure indices, 5th edition. Cincinnati, OH: American Conference of Governmental Industrial Hygienists.
2. Procter NH, Hughes JP, Fischman ML [1988]. Chemical hazards of the work place. 2nd ed. Philadelphia: J.B. Lippincott Co. Proctor and Hughes
3. ILO [1983]. Encyclopaedia of occupational health and safety, 3rd revised edition. Volumes 1 and 2. Geneva, Switzerland: International Labour Office.
4. Patty's Industrial Hygiene and Toxicology [1982]. Volumes 2A, 2B, and 2C - Toxicology. John Wiley and Sons, New York.

Table 4
Estimated Metabolic Rate - Final Finish Operator
HETA 90-246
Amfuel, Magnolia, Arkansas

	Range (Kcal/min)[a]	Estimate (Kcal/min)
Body Position 1. Sitting and standing, with some walking	0.6 to 3.0	1.5
Type of Work 1. Final Finishing - cleaning exterior of fuel cell *(Heavy work, both arms and moderate work, whole body)* 2. Final Finishing - cleaning interior of fuel cell *(Heavy work, both arms and heavy work, whole body)*	1.0 to 9.0	5.5[b]
Basal Metabolism	1.0	1.0
Summation		8.0
Hourly Estimation		480
Metabolic Rate Work Category		High

NOTES:
a   kcal/min = kilocalories per minute.
b   Metabolic estimate for cleaning *exterior* surfaces of fuel cells is 4.0; metabolic estimate for the *interior* cleaning is 7.0. Average estimate (including both interior and exterior) is 5.5.

TABLE 5
CONFINED SPACE CLASSIFICATION TABLE
HETA 90-246
Amfuel, Magnolia, Arkansas

Parameters	Class A	Class B	Class C
Characteristics	Immediately dangerous to life - rescue procedures require the entry of more than one individual fully equipped with life support equipment - maintenance of communication requires an additional standby person stationed within the confined space	Dangerous, but not immediately life threatening - rescue procedures require the entry of no more than one individual fully equipped with life support equipment - indirect visual or auditory communication with workers	Potential hazard - requires no modification of work procedures - standard rescue procedures - direct communication with workers, from outside the confined space
Oxygen	16% or less *(122 mm Hg) or greater than 25% *(190 mm HG)	16.1% to 19.4% *(122 - 147 mm Hg) or 21.5% to 25% (163 - 190 mm Hg)	19.5 % - 21.4% *(148 - 163 mm Hg)
Flammability Characteristics	20% or greater of LFL	10% - 19% LFL	10% LFL or less
Toxicity	**IDLH	greater than contamination level, referenced in 29 CFR Part 1910 Sub Part Z - less than **IDLH	less than contamination level referenced in 29 CFR Part 1910 Sub Part Z

* Based upon a total atmospheric pressure of 760 mm Hg (sea level)
** Immediately Dangerous to Life or Health - as referenced in NIOSH Registry of Toxic and Chemical Substances, Manufacturing Chemists data sheets, industrial hygiene guides or other recognized authorities.

TABLE 6
CHECK LIST OF CONSIDERATIONS FOR ENTRY,
WORKING IN AND EXITING CONFINED SPACES
HETA 90-246
Amfuel, Magnolia, Arkansas

	ITEM	CLASS A	CLASS B	CLASS C
1.	Permit	X	X	X
2.	Atmospheric Testing	X	X	X
3.	Monitoring	X	O	O
4.	Medical Surveillance	X	X	O
5.	Training of Personnel	X	X	X
6.	Labeling and Posting	X	X	X
7.	Preparation			
	Isolate/lockout/tag	X	X	O
	Purge and ventilate	X	X	O
	Cleaning Processes	O	O	O
	Requirements for special equipment/tools	X	X	O
8.	Procedures			
	Initial plan	X	X	X
	Standby	X	X	O
	Communications/observation	X	X	X
	Rescue	X	X	X
	Work	X	X	X
9.	Safety Equipment and Clothing			
	Head protection	O	O	O
	Hearing protection	O	O	O
	Hand protection	O	O	O
	Foot protection	O	O	O
	Body protection	O	O	O
	Respiratory protection	O	O	X
	Safety belts	X	X	
	Life lines, harness	X	O	
10.	Rescue Equipment	X	X	X
11.	Recordkeeping/Exposure	X	X	

X = indicates requirement
O = indicates determination by the qualified person

197

TABLE 7
SAMPLING AND ANALYTICAL METHODS
HETA 90-246
AMFUEL, MAGNOLIA, ARKANSAS

METHOD (where applicable)	COLLECTION DEVICE	SAMPLING FLOW RATE	ANALYTICAL METHOD	COMMENTS
NIOSH Method No. 2500 (Methyl ethyl ketone) (MEK)	ORBO® 90 adsorbent tubes	20 cc/min (full-shift samples) 50 cc/min (short-term samples) 100 cc/min (short-term samples)	Gas chromatography, flame ionization detector Limit of Detection = 0.01 mg/sample Limit of Quantitation = 0.03 mg/sample	Short-term (15 to 20 minutes) samples collected to evaluate "peak" exposures during the work day.
NIOSH Method No. 1003 (1,1,1-trichloroethane) (methyl chloroform)	Charcoal tubes (100/50 mg size)	20 cc/min (full-shift samples) 50 cc/min (short-term samples)	Gas chromatography, flame ionization detector Limit of Detection = 0.01 mg/sample Limit of Quantitation = 0.03 mg/sample	Short-term (15 to 20 minutes) samples collected to evaluate "peak" exposures during the work day.
HEAT STRESS	Wibget® direct reading wet bulb globe temperature monitor with data logging capability	Temperature readings recorded every 5 minutes in areas evaluated	Direct reading instrument. $WBGT_{in} = 0.7$ (Wet bulb) $+ 0.3$ (Globe) $WBGT_{out} = 0.7$ (Wet bulb) $+ 0.2$ (Globe) $+ 0.1$ (Dry bulb)	General area measurements collected throughout the work day and compared to NIOSH and ACGIH heat stress criteria.

Abbreviations: cc/min = cubic centimeters of air per minute
FID = Flame Ionization Detector
WB = Wet Bulb Temperature
Globe = Globe Temperature

GC = Gas Chromatography
WBGT = **Wet Bulb Globe Temperature**
DB = Dry Bulb Temperature

Source for analytical methods:

Eller PM, ed. [1989]. NIOSH manual of analytical methods. 3rd rev. ed. Cincinnati, OH: U.S. Department of Health and Human Services, Public Health Service, Centers for Disease Control, National Institute for Occupational Safety and Health, DHHS (NIOSH)

TABLE 8
RESULTS FROM PERSONAL BREATHING-ZONE AND GENERAL AREA AIR SAMPLES FOR METHYL ETHYL KETONE
HETA 90-246
Amfuel, Magnolia, Arkansas

DATE	SAMPLE No.	Sample Type	DEPARTMENT	ACTIVITY	TIME PERIOD SAMPLED	SAMPLE VOLUME (LITERS)	CONCENTRATION (PPM)
8/20/91	1	PBZ	Final Finish	KC-135. Wiping outside with MEK for 15 minutes. Table #26	7:48 am to 8:06 am	0.9	68
8/20/91	8	PBZ	Final Finish	KC-135. Wiping outside of cell (forward portion). Table #26	10:05 am to 10:21 am	0.8	51
8/20/91	10	PBZ	Final Finish	F-18. First cleaning. Table #2	8:30 am to 10:36 am	2.5	42
8/20/91	12	PBZ	Final Finish	KC-135. Wiping outside of cell (forward portion). Table #26	10:21 am to 11:00 am	1.0	78
8/20/91	13	PBZ	Final Finish	KC-135. Completed outside cleaning of cell. Began suspending cell and cleaning inside.	11:00 am to 11:15 am	0.8	12
8/20/91	15	PBZ	Final Finish	F-16 (A model). Begin outside cleaning. Table #16	9:18 am to 11:20 am	2.4	82
8/20/91	25	PBZ	Final Finish	KC-135. Wiping outside with MEK for 15 minutes. Table #26	10:02 am to 1:16 pm	3.9	41
8/20/91	26	AREA	Final Finish	On Table #13	9:42 am to 1:20 pm	10.9	7
8/20/91	28	PBZ	Final Finish	F-18. First cleaning. Began inside cleaning at 2:05 pm. Table #2	10:36 am to 2:02 pm	4.1	47
8/20/91	30	AREA	Final Finish	On Table #13	1:20 pm to 3:18 pm	5.9	10
8/20/91	32	PBZ	Final Finish	F-16 (A model). Continuation of outside cleaning. Table #16	11:20 am to 3:42 pm	5.4	18
8/20/91	35	PBZ	Final Finish	KC-135. Wiping outside of cell (forward portion). Table #26	1:16 pm to 3:49 pm	3.1	37

TABLE 8
RESULTS FROM PERSONAL BREATHING-ZONE AND GENERAL AREA AIR SAMPLES FOR METHYL ETHYL KETONE
HETA 90-246
Amfuel, Magnolia, Arkansas

DATE	SAMPLE No.	Sample Type	DEPARTMENT	ACTIVITY	TIME PERIOD SAMPLED	SAMPLE VOLUME (LITERS)	CONCENTRATION (PPM)
8/20/91	36	PBZ	Final Finish	F-18. Continued inside cleaning. Table #2	2:02 pm to 3:55 pm	2.3	162
8/20/91	37	PBZ	Cement House	Brush cleaning. Short-term activity performed on 2nd shift. Short-term sample.	3:29 pm to 3:48 pm	0.9	117
8/20/91	40	PBZ	Final Finish	KC-135. Beginning to wipe exterior of cell. Table #26	2:46 pm to 4:16 pm	4.5	29
8/20/91	43	PBZ	Old Cement House	Pot washing. Operation performed on 3rd shift. Consecutive short-term samples collected.	11:45 pm to 12:05 am	1.0	37
8/21/91	44	PBZ	Old Cement House	Pot washing. Operation performed on 3rd shift. Consecutive short-term samples collected.	12:05 am to 12:20 am	0.8	38
8/21/91	45	PBZ	Old Cement House	Pot washing. Operation performed on 3rd shift. Consecutive short-term samples collected.	12:20 am to 12:35 am	0.8	50
8/21/91	46	PBZ	Old Cement House	Pot washing. Operation performed on 3rd shift. Consecutive short-term samples collected.	12:35 am to 12:50 am	0.8	44
8/21/91	47	PBZ	Face Coater	Face coating end. Cleaning roller and changing cements (#899 to PU 257). Cleaning and change-over completed at 8:55 am.	8:42 am to 8:57 am	0.8	131
8/21/91	48	PBZ	Cement Spray Room	Changing cements (to PF149, which contains MEK and toluene). Using methyl n-propyl ketone. Continued spraying after changing cements.	9:15 am to 9:30 am	0.8	9
8/21/91	52	PBZ	Face Coater	Take up end (opposite of face coating end).	7:48 am to 10:20 am	3.0	48

TABLE 8
RESULTS FROM PERSONAL BREATHING-ZONE AND GENERAL AREA AIR SAMPLES FOR METHYL ETHYL KETONE
HETA 90-246
Amfuel, Magnolia, Arkansas

DATE	SAMPLE No.	Sample Type	DEPARTMENT	ACTIVITY	TIME PERIOD SAMPLED	SAMPLE VOLUME (LITERS)	CONCENTRATION (PPM)
8/21/91	53	PBZ	Face Coater	Face coater end.	7:50 am to 10:22 am	3.0	45
8/21/91	54	PBZ	Cement Spray Room	Cement spraying.	7:37 am to 10:40 am	3.7	6
8/21/91	55	PBZ	"New" Cement Spray Room	Alternates spraying MEK-based cements and latex coatings. Two different spray booths.	8:22 am to 10:37 am	2.7	6
8/21/91	58	AREA	Cement Spray Room	Spraying MEK-based cements.	8:33 am to 11:33 am	3.6	3
8/21/91	59	PBZ	Cement House	Cement mixer. Mixing a cement batch during this short-term sample.	12:36 pm to 12:51 pm	1.5	61
8/21/91	60	PBZ	Cement House	Cement mixer. Mixing a cement batch during this short-term sample. Dispensing MEK to a 55-gallon container from bulk storage using a nozzle dispenser.	12:51 pm to 1:06 pm	1.5	32
8/21/91	61	PBZ	Face Coater	Take up end (opposite of face coating end).	10:20 am to 1:20 pm	3.6	24
8/21/91	62	PBZ	Face Coater	Face coating end.	10:22 am to 1:22 pm	3.6	18
8/21/91	63	PBZ	Cement House	Cement mixer. Dispensing MEK to a 55-gallon container from bulk storage using a nozzle dispenser.	1:07 pm to 1:23 pm	1.6	44
8/21/91	64	PBZ	Face Coater	Transfer (by hand) of PU-267 cement from a 5 gallon bucket to the face coating tray using a cardboard tub.	2:09 pm to 2:24 pm	0.8	5
8/21/91	65	PBZ	Fittings	Ring cleaning.	2:24 pm to 2:39 pm	0.8	149

TABLE 8
RESULTS FROM PERSONAL BREATHING-ZONE AND GENERAL AREA AIR SAMPLES FOR METHYL ETHYL KETONE
HETA 90-246
Amfuel, Magnolia, Arkansas

DATE	SAMPLE No.	Sample Type	DEPARTMENT	ACTIVITY	TIME PERIOD SAMPLED	SAMPLE VOLUME (LITERS)	CONCENTRATION (PPM)
8/21/91	66	PBZ	Fittings	Ring cleaning.	2:39 pm to 2:54 pm	0.87	276
8/21/91	67	PBZ	Face Coater	Transfer (by hand) of PU-267 cement from a 5 gallon bucket to the face coating tray using a cardboard tub.	2:45 pm to 2:58 pm	0.7	89
8/21/91	68	PBZ	Face Coater	Take up end (opposite of face coating end).	1:20 pm to 3:00 pm	2.0	11
8/21/91	69	PBZ	Face Coater	Face coating end.	1:22 pm to 3:01 pm	2.0	24
8/21/91	70	PBZ	Fittings	Ring cleaning (employee left for break during this sampling period so ring washing activity was interrupted).	2:54 pm to 3:29 pm	1.8	115
8/21/91	71	PBZ	Fittings	Ring cleaning performed throughout this sampling period.	3:29 pm to 3:44 pm	0.8	421
8/21/91	72	PBZ	Fittings	Ring cleaning performed throughout this sampling period.	3:44 pm to 3:59 pm	0.8	326
8/21/91	75	PBZ	Fittings	Full-shift sample collected on employee assigned to ring cleaning. Also performed other duties which did not involve solvents.	8:10 am to 4:00 pm	9.4	47
8/21/91	92	PBZ	"New" Cement Spray Room	Alternates spraying MEK-based cements and latex coatings. Uses two different spray booths.	10:37 am to 1:19 pm	3.2	7
8/21/91	93	PBZ	Cement Spray Room	Spraying MEK-based cements.	10:40 am to 1:30 pm	3.4	5
8/21/91	95	PBZ	Cement Spray Room	Spraying MEK-based cements.	1:30 pm to 2:45 pm	1.5	12

TABLE 8
RESULTS FROM PERSONAL BREATHING-ZONE AND GENERAL AREA AIR SAMPLES FOR METHYL ETHYL KETONE
HETA 90-246
Amfuel, Magnolia, Arkansas

DATE	SAMPLE No.	Sample Type	DEPARTMENT	ACTIVITY	TIME PERIOD SAMPLED	SAMPLE VOLUME (LITERS)	CONCENTRATION (PPM)
8/21/91	97	PBZ	"New" Cement Spray Room	Alternates spraying MEK-based cements and latex coatings. Uses two different spray booths.	1:19 pm to 2:51 pm	1.8	7
8/21/91	98	AREA	Cement Spray Room	Area sample collected within the cement spray room	11:42 am to 2:42 pm	3.6	2
			Evaluation Criteria	NIOSH Recommended Exposure Limit			100 ppm, 8-hour TWA 300 ppm, STEL
				OSHA Permissible Exposure Limit			100 ppm, 8-hour TWA 300 ppm, STEL
				ACGIH Threshold Limit Value			100 ppm, 8-hour TWA 300 ppm, STEL

Comments:

1. TWA = time weighted average
2. STEL = 15-minute short-term exposure level

TABLE 9
RESULTS FROM PERSONAL BREATHING ZONE AND GENERAL AREA AIR SAMPLES FOR 1,1,1-trichloroethane
HETA 90-246
Amfuel, Magnolia, Arkansas

DATE	SAMPLE No.	SAMPLE TYPE	DEPARTMENT	ACTIVITY	TIME PERIOD SAMPLED	SAMPLE VOLUME (LITERS)	CONCENTRATION (PPM)
8/20/91	2	PBZ	Final Finish	F-16A. Cleaning the interior of the cell. Two blowers in use. Two employees inside cell. Table #12	8:03 am to 8:19 am	0.8	412
8/20/91	3	PBZ	Final Finish	F-16A. Cleaning the interior of the cell. Two blowers in use. Two employees inside cell. Table #12	8:19 am to 8:36 am	0.9	560
8/20/91	5	PBZ	Final Finish	F-16A. Finished cleaning the interior of the cell at 9:05 am. Two employees inside cell. Table #12	8:36 am to 9:10 am	1.7	312
8/20/91	6	PBZ	Final Finish	KC-135. Wiping and painting the exterior of the cell (forward section). Two employees. Table #26	7:29 am to 9:24 am	2.3	318
8/20/91	7	PBZ	Final Finish	A6 (Aft fuel cell). Working inside the fuel cell during this sampling period.	7:29 am to 9:23 am	2.3	422
8/20/91	9	PBZ	Final Finish	KC-135. Interior cleaning until approximately 9:30 am. Table #22	7:48 am to 9:32 am	2.1	383
8/20/91	11	PBZ	Final Finish	F-18 (#2 cell). Working on the exterior/interior of the cell. Table #20	10:36 am to 10:55 am	1.0	674
8/20/91	16	PBZ	Final Finish	F-18 (#2 cell). Working on the exterior/interior of the cell. Table #20	10:55 am to 11:25 am	1.5	793
8/20/91	17	PBZ	Final Finish	F-16 (A model). Cleaning outside of the cell. Table #16	11:25 am to 11:40 am	0.8	878
8/20/91	18	PBZ	Final Finish	F-16 (A model). Cleaning outside of the cell. Table #16	11:40 am to 11:55 am	0.8	293
8/20/91	19	PBZ	Final Finish	A6 (aft cell). Cleaning the interior of the fuel cell. Table #5	9:24 am to 12:42 pm	4.0	92

TABLE 9
RESULTS FROM PERSONAL BREATHING ZONE AND GENERAL AREA AIR SAMPLES FOR 1,1,1-trichloroethane
HETA 90-246
Amfuel, Magnolia, Arkansas

DATE	SAMPLE No.	SAMPLE TYPE	DEPARTMENT	ACTIVITY	TIME PERIOD SAMPLED	SAMPLE VOLUME (LITERS)	CONCENTRATION (PPM)
8/20/91	20	PBZ	Final Finish	KC-135. Interior cleaning until approximately 9:30 am. On break from 9:32 am to 9:45 am. Table #22	9:46 am to 12:55 pm	3.8	53
8/20/91	21	Area	Final Finish	On Table #13	9:42 am to 12:59 pm	9.9	5
8/20/91	22	PBZ	Final Finish	A6 (Aft fuel cell) and F-15 cell. Involved in cleaning both cells during this sampling period.	9:23 am to 1:03 pm	4.4	83
8/20/91	24	PBZ	Final Finish	KC-135. Cleaning the interior (forward) portion of the fuel cell. Two workers assigned to cell. Table #26.	12:44 pm to 1:12 pm	1.4	405
8/20/91	27	PBZ	Final Finish	KC-135. Cleaning the interior (forward) portion of the fuel cell. A series of consecutive samples were collected at this location. Table #26.	1:12 pm to 1:58 pm	2.3	573
8/20/91	29	PBZ	Final Finish	KC-135. Finished cleaning the interior (forward) portion of the fuel cell around 2:30 pm. Table #26.	1:58 pm to 2:35 pm	1.9	3
8/20/91	31	Area	Final Finish	On Table #13.	12:59 pm to 3:18 pm	7.0	10
8/20/91	33	PBZ	Final Finish	F-16(A). Cleaning the interior of the fuel cell. Table #12.	1:47 pm to 3:42 pm	2.3	469
8/20/91	34	PBZ	Final Finish	KC-135. Cleaning the interior (forward) portion of the fuel cell. Table #26	12:42 pm to 3:49 pm	3.7	134
8/20/91	38	PBZ	Final Finish	F-18 (#4 cell). Stenciling inside the fuel cell. No respirator worn.	12:55 pm to 4:06 pm	3.8	25

TABLE 9
RESULTS FROM PERSONAL BREATHING ZONE AND GENERAL AREA AIR SAMPLES FOR 1,1,1-trichloroethane
HETA 90-246
Amfuel, Magnolia, Arkansas

DATE	SAMPLE No.	SAMPLE TYPE	DEPARTMENT	ACTIVITY	TIME PERIOD SAMPLED	SAMPLE VOLUME (LITERS)	CONCENTRATION (PPM)
8/20/91	39	PBZ	Final Finish	F-15 (300-1). Cleaning the interior of the cell. Employee placed head and arms inside the cell during cleaning process. No respirator worn by the worker during this cleaning process since only their head was inside the fuel cell.	1:02 pm to 4:14 pm	3.8	47
8/20/91	41	PBZ	Final Finish	Table #26. Finished cleaning of a KC-135 cell at 2:30 pm	2:35 pm to 4:16 pm	5.1	161
	Evaluation Criteria		NIOSH Recommended Exposure Limit				350 ppm CL*
			OSHA Permissible Exposure Limit				350 ppm TWA 450 ppm STEL
			ACGIH Threshold Limit Value				350 ppm TWA 450 ppm STEL

TABLE 10
RESULTS FROM AIR SAMPLES FOR ETHANOL
HETA 90-246
Amfuel, Magnolia, Arkansas

DATE	SAMPLE No.	OPERATION	TIME PERIOD SAMPLED	SAMPLE VOLUME (LITERS)	CONCENTRATION (PPM)
8/21/91	49	Nylon sprayer	0739 to 0945	6.3	35
8/21/91	56	Nylon sprayer	0945 to 1102	3.9	26
8/21/91	94	Nylon sprayer	1105 to 1616	15.6	160

Evaluation Criteria	NIOSH Recommended Exposure Limit	1000, 8-hour TWA
	OSHA Permissible Exposure Limit	1000, 8-hour TWA
	ACGIH Threshold Limit Value	1000, 8-hour TWA

TABLE 11
Results of Heat Stress Monitoring (°F)
American Fuel Cell and Coated Fabrics Company
Magnolia, Arkansas
August 20, 1991
HETA 90-246

Time	Wet Bulb	Dry Bulb	Globe Temp.	WBGT$_{in}$	WBGT$_{out}$
System 2 (Uniroyal)					
0940	72.9	89.4	90.4	78.1	----
1050	71.9	91.1	92.3	78.1	----
1240	73.0	95.3	96.4	80.0	----
1400	74.4	97.0	98.2	81.5	----
Finish Area, Table #5 (outside cell)					
0935	72.6	89.1	90.0	77.9	----
1045	71.4	89.5	90.3	77.1	----
1230	72.0	93.4	94.2	78.6	----
1355	73.7	95.8	96.3	80.5	----
Finish Area, Table #5 (inside cell)					
0930	73.7	89.4	90.0	78.6	----
Outerply (old building)					
0945	71.3	87.6	88.2	76.6	----
1105	71.3	90.6	90.1	77.0	----
1245	72.4	93.6	94.7	79.0	----
1405	73.4	95.0	96.4	80.2	----
Inner Liner (new building, near drying table)					
0955	70.9	87.2	88.1	76.0	----
1115	70.7	90.3	90.1	76.5	----
1255	71.6	93.4	94.1	78.3	----
1415	73.1	96.2	96.8	80.2	----
Tab Area (near pre-shrink oven)					
1000	71.9	89.3	90.2	77.2	----
1120	70.9	90.4	90.9	76.9	----
1300	72.6	96.0	97.3	79.8	----
1425	73.9	98.0	99.0	81.3	----
Fittings Department (buffing and finishing)					
1010	70.0	84.9	85.9	74.9	----
1125	69.9	87.2	88.4	75.4	----
1305	70.9	90.6	92.3	77.5	----
1430	72.4	91.7	94.2	78.9	----
Fittings Department (press area)					
1015	71.6	91.1	91.3	77.6	----
1130	71.6	94.9	96.1	79.1	----
1310	73.4	96.8	96.7	80.3	----
1435	75.1	100.7	101.4	83.0	----
Face Coating Department					
1020	71.2	89.7	92.0	77.4	----
1140	72.4	93.4	96.3	79.5	----
1320	74.6	98.4	100.0	82.4	----
1445	73.8	96.5	100.3	81.7	----
Outside (between new building and cement house)					
1005	70.6	86.0	108.2	----	79.6
1145	72.8	92.6	120.5	----	83.8
1325	74.4	96.8	122.0	----	86.3
1450	74.4	96.4	116.4	----	84.8

Table 12

Results From Personal Heat Stress Dosimetry
Amfuel, Magnolia, Arkansas
HETA 90-246

Time	Ear Temperature		Mold Temperature	
	°C	°F	°C	°F
07:19	34.8	94.6	30.8	87.4
07:24	36.8	98.3	30.7	87.2
07:29	36.9	98.4	30.5	86.9
07:34	36.8	98.3	29.9	85.8
07:39	37.0	98.6	30.2	86.3
07:44	37.0	98.6	30.2	86.4
07:49	36.9	98.4	29.9	85.8
07:54	36.9	98.4	29.6	85.3
07:59	36.9	98.4	29.8	85.7
08:04	36.9	98.4	30.0	86.0
08:09	36.8	98.3	30.0	86.0
08:14	36.8	98.3	30.0	86.0
08:19	36.7	98.1	30.0	85.9
08:24	36.8	98.3	30.0	86.0
08:29	36.7	98.1	30.1	86.1
08:34	36.7	98.1	30.4	86.7
08:39	36.8	98.3	30.8	87.4
08:44	36.7	98.1	31.2	88.1
08:49	36.8	98.3	31.3	88.3
08:54	36.8	98.3	31.5	88.7
08:59	36.9	98.4	31.8	89.2
09:04	37.0	98.6	32.1	89.8
09:09	37.0	98.6	32.2	90.0
09:14	37.1	98.8	32.4	90.4
09:19	37.1	98.8	32.6	90.8
09:24	37.1	98.8	32.6	90.8
09:29	37.1	98.8	32.7	90.9
09:34	36.8	98.3	31.9	89.4
09:39	36.7	98.1	29.1	84.4
09:44	36.8	98.3	30.2	86.4
09:49	36.8	98.3	31.5	88.7
09:54	36.9	98.4	31.8	89.2

Comments:

Start Time: 7:19 am
End Time: 9:58 am
Total Run Time: 2:38:40
Alarm Level Setting: 39.0°C
Sample Rate: 5 minutes
High Temperature: 37.2°C (at 9:22 am)
Low Temperature: 34.7°C (at 9:58 am)

Appendix A
Elements of a Comprehensive Heat Stress Management Program
HETA 90-246
Amfuel, Magnolia, Arkansas

1. **Written program** - A detailed written document is necessary to specifically describe the company procedures and policies in regards to heat management. The input from management, technical experts, physician(s), labor union, <u>and</u> the affected employees should be considered when developing the heat management program. This program can only be effective with the full support of plant management.

2. **Environmental monitoring** - In order to determine which employees should be included in the heat management program, monitoring the environmental conditions is essential. Environmental monitoring also allows one to determine the severity of the heat stress potential during normal operations and during heat alert periods.

3. **Medical examinations and policies** - Preplacement and periodic medical examinations should be provided to <u>all</u> employees included in the heat management program where the work load is heavy or the environmental exposures are extreme. Periodic exams should be conducted at least annually, ideally immediately prior to the hot season (if applicable). The examination should include a comprehensive work and medical history with special emphasis on any suspected previous heat illness or intolerance. Organ systems of particular concern include the skin, liver, kidney, nervous, respiratory, and circulatory systems. Written medical policies should be established which clearly describe specific predisposing conditions that cause the employee to be at higher risk of a heat stress disorder, and the limitations and/or protective measures implemented in such cases.

4. **Work schedule modifications** - The work-rest regime can be altered to reduce the heat stress potential. Shortening the duration of work in the heat exposure area and utilizing more frequent rest periods reduces heat stress by decreasing the metabolic heat production and by providing additional recovery time for excessive body heat to dissipate. Naturally, rest periods should be spent in cool locations (preferably air conditioned spaces) with sufficient air movement for the most effective cooling. Allowing the worker to self-limit their exposure on the basis of signs and symptoms of heat strain is especially protective since the worker is usually capable of determining their individual tolerance to heat. However, there is a danger that under certain conditions, a worker may not exercise proper judgement and experience a heat-induced illness or accident.

5. **Acclimatization** - Acclimatization refers to a series of physiological and psychological adjustments that occur which allow one to have increased heat tolerance after continued and prolonged exposure to hot environmental conditions. Special attention must be given when administering work schedules during the beginning of the heat season, after long weekends or vacations, for new or temporary employees, or for those workers who may otherwise be unacclimatized because of their increased risk of a heat-induced accident or illness. These employees should have reduced work loads (and heat exposure durations) which are gradually increased until acclimatization has been achieved (usually within 4 or 5 days).

6. **Clothing** - Clothing can be used to control heat stress. Workers should wear clothing which permits maximum evaporation of perspiration, and a minimum of perspiration run-off which does not provide heat loss, (although it still depletes the body of salt and water). For extreme conditions, the use of personal protective clothing such as a radiant reflective clothing, and torso cooling vests should be considered.

7. **Buddy system** - No worker should be allowed to work in designated hot areas without another person present. A buddy system allows workers to observe fellow workers during their normal job duties for early signs and symptoms of heat intolerance such as weakness, unsteady gait, irritability, disorientation, skin color changes, or general malaise, and would provide a quicker response to a heat-induced incident.

8. **Drinking water** - An adequate amount of cool (50-60°F) potable water should be supplied within the immediate vicinity of the heat exposure area as well as the resting location(s). Workers who are exposed to hot environments are encouraged to drink a cup (approximately 5-7 ounces) every 15-20 minutes even in the absence of thirst.

9. **Posting** - Dangerous heat stress areas (especially those requiring the use of personal protective clothing or equipment) should be posted in readily visible locations along the perimeter entrances. The information on the warning sign should include the hazardous effects of heat stress, the required protective gear for entry, and the emergency measures for addressing a heat disorder.

10. **Heat alert policies** - A heat alert policy should be implemented which may impose restrictions on exposure durations (or otherwise control heat exposure) when the National Weather Service forecasts that a heat wave is likely to occur. A heat wave is indicated when daily maximum temperature exceeds 95°F or when the daily maximum temperature exceeds 90°F and is at least 9°F more than the maximum reached on the preceding days.

11. **Emergency contingency procedures** - Well planned contingency procedures should be established in writing and followed during times of a

heat stress emergency. These procedures should address initial rescue efforts, first aid procedures, victim transport, medical facility/service arrangements, and emergency contacts. Specific individuals (and alternatives) should be assigned a function within the scope of the contingency plan. Everyone involved must memorize their role and responsibilities since response time is critical during a heat stress emergency.

12. **Employee education and training** - All employees included in the heat management program or emergency contingency procedures should receive periodic training regarding the hazards of heat stress, signs and symptoms of heat-induce illnesses, first aid procedures, precautionary measures, and other details of the heat management program.

13. **Assessment of program performance and surveillance of heat-induced incidents** - In order to identify deficiencies with the heat management program a periodic review is warranted. Input from the workers affected by the program is necessary for the evaluation of the program to be effective. Identification and analysis of the circumstances pertinent to any heat-induce accident or illness is also crucial for correcting program deficiencies.

Appendix B
Selected Local Exhaust Ventilation Designs

Source:
Industrial Ventilation Manual, 19th Edition
American Conference of Governmental Industrial Hygienists

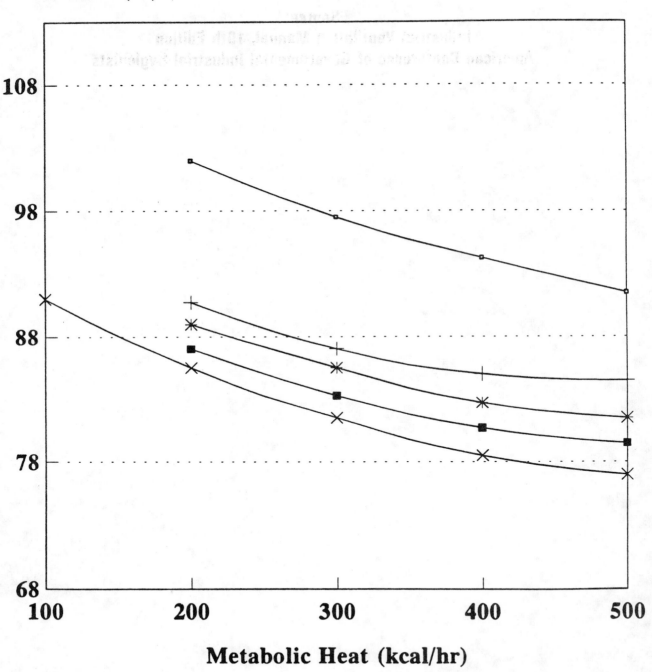

Figure 1 - NIOSH Recommended Exposure Limits
Heat Acclimatized Workers

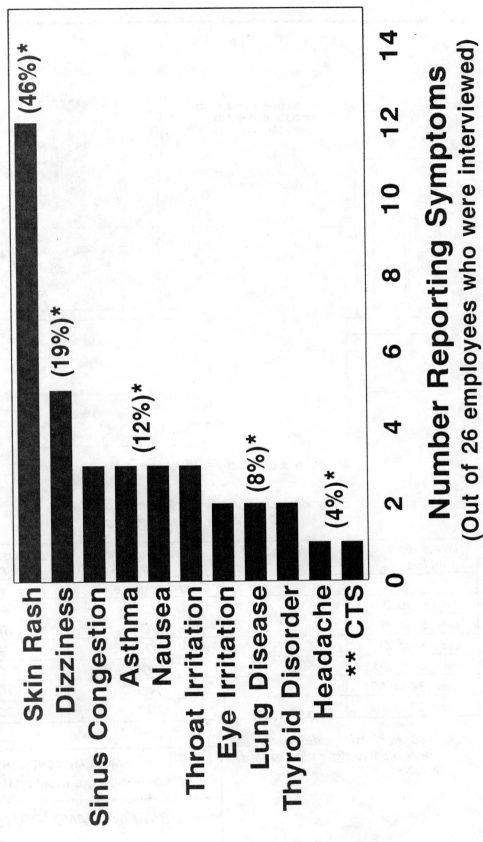

Figure 2: Medical Interview Data
Amfuel Corporation, HETA #90-246
November 14, 1990

Comments:
* Percent of the 26 interviewed who reported the symptom
** Carpal Tunnel Syndrome

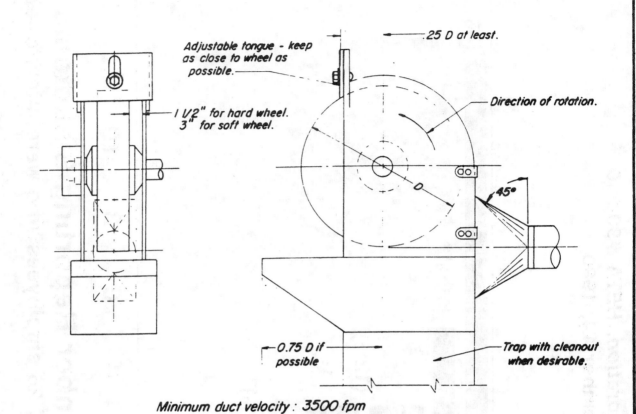

Minimum duct velocity: 3500 fpm

Entry loss: 0.65 VP for straight take-off.
0.40 VP for tapered take-off.

Wheel diam. inches	Wheel width * inches	Exhaust volume cfm Good enclosure	Exhaust volume cfm Poor enclosure
to 9	2	300	400
over 9 to 16	3	500	610
over 16 to 19	4	610	740
over 19 to 24	5	740	1200
over 24 to 30	6	1040	1500
over 30 to 36	6	1200	1990

* In cases of extra wide wheels, use wheel width to determine exhaust volume.

AMERICAN CONFERENCE OF GOVERNMENTAL INDUSTRIAL HYGIENISTS

BUFFING AND POLISHING

DATE 1-82   VS-406

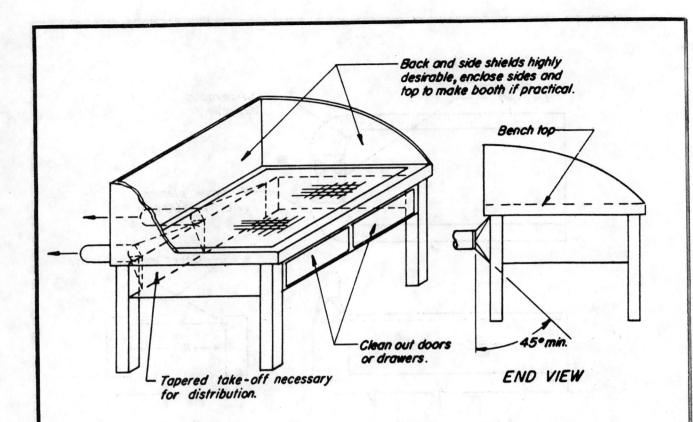

Q = 150 - 250 cfm/sq ft of bench area.
Minimum duct velocity = 3500 fpm
Entry loss = 0.25 VP for tapered take-off.

Grinding in booth, 100 fpm face velocity also suitable.

For downdraft grilles in floor: Q = 100 cfm/sq ft of working area.

Provide equal distribution. Provide for cleanout.

AMERICAN CONFERENCE OF GOVERNMENTAL INDUSTRIAL HYGIENISTS	
PORTABLE HAND GRINDING	
DATE 1-64	VS-412

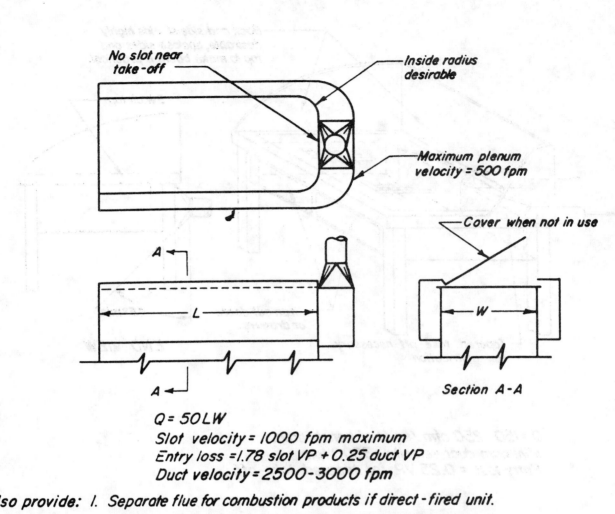

Q = 50LW
Slot velocity = 1000 fpm maximum
Entry loss = 1.78 slot VP + 0.25 duct VP
Duct velocity = 2500-3000 fpm

Also provide: 1. Separate flue for combustion products if direct-fired unit.
2. For cleaning operation, an air-line respirator is necessary.
3. For pit units, the pit should be mechanically ventilated.
4. For further safe guards, see VS-501.1
NOTE: Provide downdraft grille for parts that cannot be removed dry; Q = 50 cfm/sq ft grille area.

AMERICAN CONFERENCE OF GOVERNMENTAL INDUSTRIAL HYGIENISTS		
SOLVENT DEGREASING TANKS		
DATE	1-78	VS-501

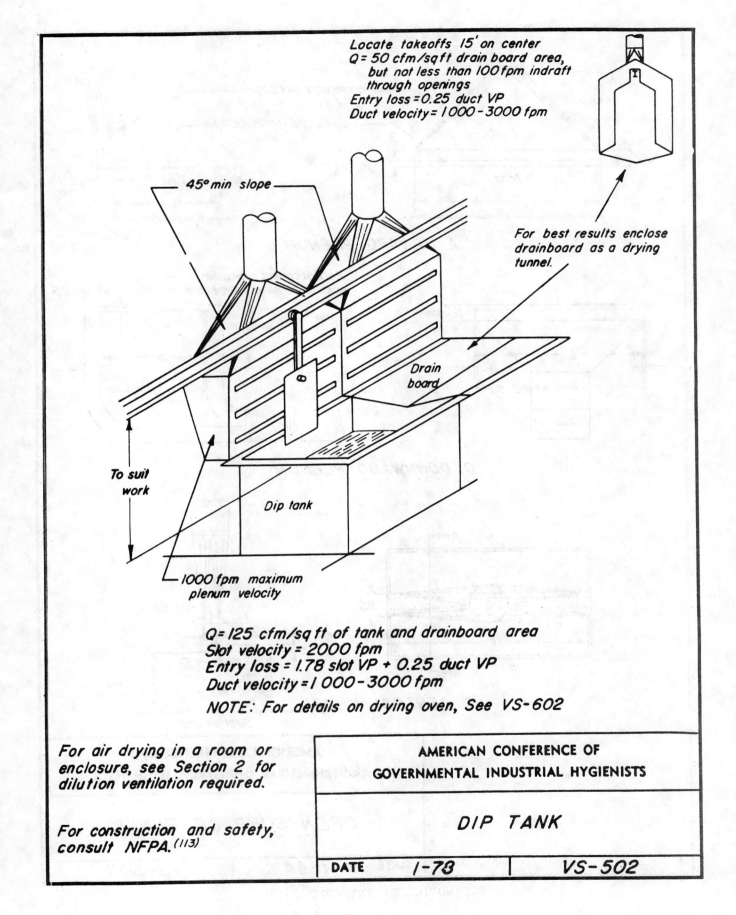

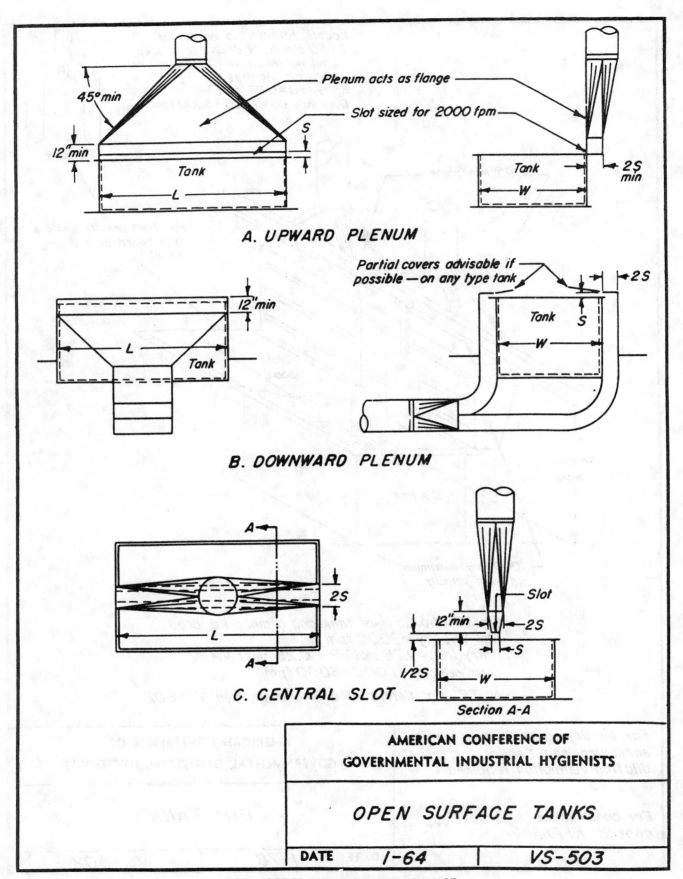

OPEN SURFACE TANKS

AMERICAN CONFERENCE OF GOVERNMENTAL INDUSTRIAL HYGIENISTS

DATE 1-64   VS-503

SEE NOTICE OF INTENDED CHANGE

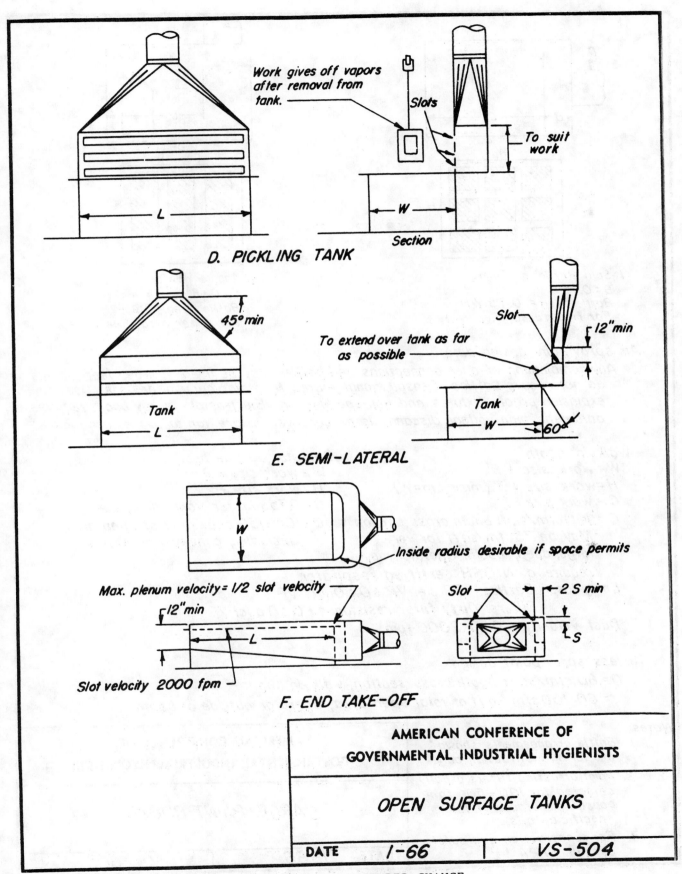

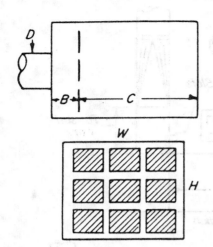 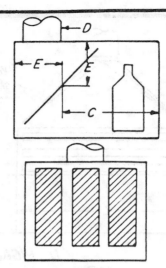

1. Split Baffle or Filters
B = 0.75 D
Baffle area = 0.75 WH
For filter area, See Note 2

2. Angular Baffle
E = D + 6"
Baffle area = 0.40 WH
For filter area, See Note 2

Air spray paint design data
    Any combination of duct connections and baffles may be used. Large, deep booths do not require baffles. Consult manufactures for water-curtain designs. Use explosion proof fixtures and non-sparking fan. Electrostatic spray booth requires automatic high-voltage disconnects for conveyor failure, fan failure or grounding.

Walk-in booth
W = work size + 6'
H = work size + 3' (minimum = 7')
C = work size + 6'
Q = 100 cfm/sq ft booth cross section
    May be 75 cfm/sq ft for very large, deep, booth. Operator may require a NIOSH certified respirator.
Entry loss = Baffles: 1.78 slot VP + 0.50 duct VP
           = Filters: Dirty filter resistance + 0.50 duct VP
Duct velocity = 1000-2000 fpm

Operator outside booth
W = work size + 2'
H = work size + 2'
C = 0.75 x larger front dimension
Q = 100-150 cfm/sq ft of open area, including conveyor openings.

Airless spray paint design
Q = 60 cfm/sq ft booth cross section, walk-in booth
   = 60-100 cfm/sq ft of total open area, operator outside of booth

Notes:
1. Baffle arrangements shown are for air distribution only.
2. Paint arresting filters usually selected for 100-500 fpm, consult manufacturer for specific details.
3. For construction and safety, consult NFPA[113]

AMERICAN CONFERENCE OF GOVERNMENTAL INDUSTRIAL HYGIENISTS	
LARGE PAINT BOOTH	
DATE 1-86	VS-603

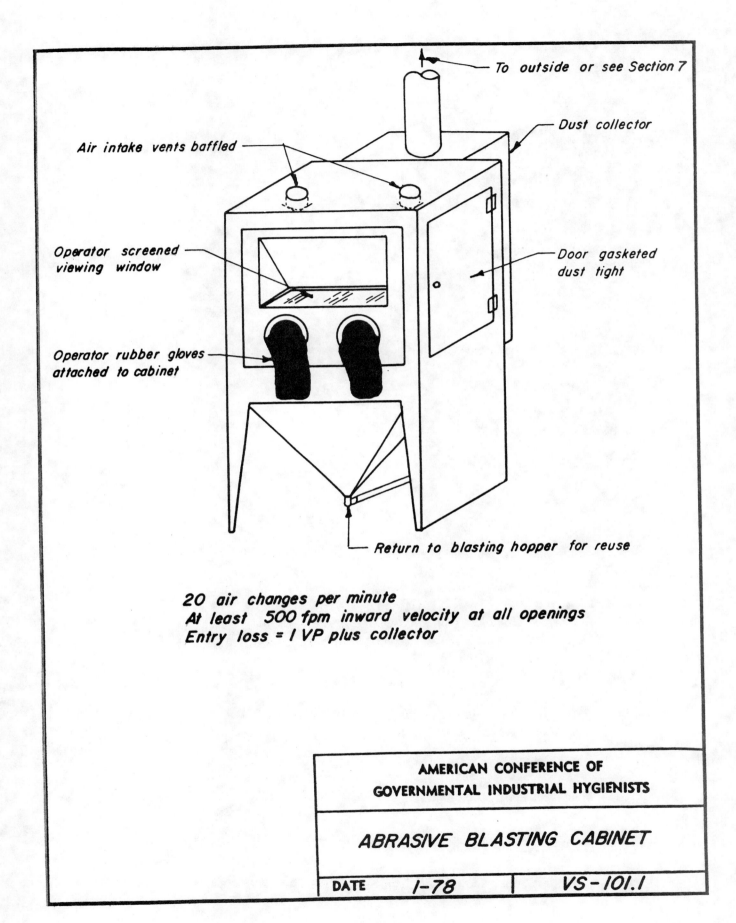

HETA 92-001-2444
AUGUST 1994
SANCAP ABRASIVES, INC.
ALLIANCE, OHIO

NIOSH INVESTIGATORS:
Faye T. Bresler, M.D., M.P.H.
Hongwei Hsiao, Ph.D.
Kevin Hanley, M.S.P.H., C.I.H.

I. SUMMARY

On September 22-23, 1992, the National Institute for Occupational Safety and Health (NIOSH) conducted a health hazard evaluation (HHE) at SANCAP Abrasives, in Alliance, Ohio. The request, submitted by the International Chemical Workers Union (ICWU), concerned musculoskeletal problems reported by AT-6 press operators. The AT-6 is a semi-automatic abrasive belt production machine that converts large sheets of abrasive paper or cloth into belts.

The objectives of the HHE were to characterize worker musculoskeletal symptoms, to identify job tasks which may cause or contribute to musculoskeletal illnesses, and propose recommendations to reduce the likelihood of developing these illnesses. The medical evaluation consisted of confidential health interviews with AT-6 workers, review of the Occupational Safety and Health Administration Log of Injuries and Illnesses (OSHA form 200), and informal discussions with the employees regarding musculoskeletal hazards associated with their jobs. The ergonomic evaluation included measurement of workstation dimensions, as well as a work methods and biomechanical analysis aided by videotaping and photography of the AT-6 process.

All ten AT-6 workers present on September 22 participated in confidential interviews. The average age was 29, with duration of employment ranging from 1-8 months. None of the current employees reported sustaining musculoskeletal illnesses on the AT-6 machine. Video analysis and process observation suggested that the major ergonomic stressors of the press operating job were associated with the upper extremities and torso, due to repetitive movements as well as postural and muscular force demands. The following risk factors for the potential development of cumulative trauma disorders were observed: repetitive 30 second task cycle; awkward and static postures; significant truncal flexion and twisting; impact force with the palm of the hand (or clinched fist); pinch grips; and extended reaches.

> Ergonomic evaluation of the AT-6 press operation found exposures to several factors, including repetition, stressful postures, and muscular force demands, that increase the risk of developing musculoskeletal disorders. Recommendations for engineering, administrative, and work practice controls are contained in Section VIII of this report.

KEYWORDS: SIC 3291 (Abrasives Products), ergonomics, cumulative trauma disorders, lumbar stress, truncal flexion, awkward and static posture, pinch grip.

Health Hazard Evaluation Report No. 92-001

II. INTRODUCTION

The National Institute for Occupational Safety and Health (NIOSH) received a request dated September 26, 1991, from the International Chemical Workers Union to conduct a health hazard evaluation (HHE) at the Swiss Industrial Abrasives/SIA America facility, now known as SANCAP Abrasives, Inc., in Alliance, Ohio. The request stated that at least three employees reported (lumbar) musculoskeletal problems due to work practices and equipment design utilized at the AT-6 press, especially during the manufacture of large-size abrasive belts.

A NIOSH site visit to the facility was scheduled for January 6-7, 1992. This site visit was postponed after NIOSH representatives were informed on January 2, 1992, that the facility was to be sold and would not be engaging in normal production the week of the planned site visit.

In February 1992, the ownership of the facility was transferred from Swiss Industrial Abrasives/SIA America, Inc., to SANCAP Abrasives, Inc. After this change in ownership, the International Chemical Workers Union no longer represented workers at the facility. The NIOSH site visit was re-scheduled for June 10-12, 1992, but was again postponed because of SANCAP's refusal to allow NIOSH investigators access to the worksite.

On September 21, 1992, NIOSH investigators obtained a research inspection warrant, and on September 22-23, 1992, conducted the HHE.

III. BACKGROUND

The AT-6 machine is located in a building, constructed in 1935, originally used during World War II for aircraft manufacturing. Since 1948, the building has been owned and operated by different corporate entities, but has consistently been used for manufacturing sandpaper and other abrasive products. In 1977, the facility was purchased by Swiss Industrial Abrasives (SIA) and was renamed SANCAP, with SIA remaining the parent company. In 1985, the facility name was changed to SIA America, with SIA still the parent company. The facility complex was comprised of two manufacturing divisions, abrasives and cap liners, located in different buildings. In 1988, SIA divested the cap liner division. In February 1992, the abrasive division was also divested and sold to the same company that had purchased the cap liner division. The new owner operates under the name SANCAP Abrasives, Incorporated. This health hazard investigation focused on the AT-6 machine located in the abrasives division.

The AT-6 machine was built in Switzerland in 1980, transported to Alliance, Ohio, and installed at the SANCAP facility. With few modifications, the machine is currently operating in the same manner as it did at the time of installation. The addition of removable side rails on the AT-6 press platforms is one modification made under previous management of the facility.

Health Hazard Evaluation Report No. 92-001

The AT-6 machine is a semi-automatic manufacturing machine that produces large industrial-sized abrasive belts which are utilized in the steel, automotive, and other industries. The belts are typically constructed of an alumina compound grit with either a paper or cloth backing. The abrasive grit is applied to one side of the backing material prior to the AT-6 process. Sheets of the abrasive paper are cut to size and hung on the transfer conveyor of the AT-6 machine. The abrasive grit is removed from two edges, an adhesive is applied, and the sheet is conveyed to the AT-6 press. Press operators remove the sheet from the conveyor, position the edges together, and activate the press to form the belt. The finished belt is removed from the press, packaged, and sent to the shipping department.

IV. WORK DESCRIPTION

The AT-6 press operation job is manually intensive work, requiring sustained eye-hand coordination. During the NIOSH site visit, operators were observed sequentially performing the following procedures: pick-up an abrasive sheet from a conveyor; carry it to, and place it on, the press; position and apply manual force to the two ends of the sheet, which have adhesive; activate the mechanical force of the press to convert the sheet to a belt; and, finally, remove any remaining adhesive from the press table or abrasive belt. Three workers were typically present during this operation during the NIOSH site visit - two press operators working with the same abrasive sheet on opposite sides of the press, and a material handler packaging the finished product. Refer to Figure 1 for diagrams of the operators' typical posture during each task of this work cycle that was observed from the videotape analysis. AT-6 press operators stand on movable platforms and lean forward to position the belt and operate the press. The length and width of the belts produced affect how the platforms are positioned at the AT-6 Press. Larger belts, especially wider belts, require the platforms to be moved farther apart to avoid damaging the product by contact with the platform.

Abrasive belts are manufactured by SANCAP according to specific customer orders, with length, width, grit type, backing (paper/cloth), and number of belts produced varying in each production run. The widths of the abrasive belts range from 15" to 64", and the lengths vary from 103" to 128". Two operators usually work together to operate the AT-6 press machine; during the production of 64" wide belts, four workers operate the press. The operators work 10 hours a day, from 6 a.m. to 4:30 p.m., four days a week. During a 10-hour shift, there are many production runs, each consisting of one customer's order. The smallest orders consist of 15-20 belts, and the largest orders as high as 250 belts. Within a given production run the manufacturing process is continuous; there is a half-hour machine "down-time" per process change to complete the machine adjustments. (The actual production time delay is dependant on the specific adjustments required between different products.) During the down-time, operators perform miscellaneous tasks. Hundreds of belts are produced daily on the AT-6 machine. The operators have one scheduled

Health Hazard Evaluation Report No. 92-001

30-minute lunch break and three scheduled 10-minute breaks during the 10 hour work day.

V. EVALUATION CRITERIA

In general, the evaluation of occupational musculoskeletal problems includes the investigation of human-machine interactions, and involves a field of study called ergonomics. Basic production information is necessary to understand the demands made on humans as they interact with machines in an industrial environment. The specific tasks performed by individuals, as well as the number of repetitive movements during a work cycle, duration of work cycles, and the rest-time between sets of repetitions, are some of the important factors in an ergonomic evaluation.

Cumulative Trauma Disorders

The occurrence of cumulative trauma disorders (CTDs) of the musculoskeletal system as a result of occupational and non-occupational factors has been reported over many years.[1,2,3,4] These musculoskeletal disorders frequently affect the tendons, tendon sheaths, muscles, and nerves of the upper extremities (UE). Common UE CTDs include tendinitis, synovitis, tenosynovitis, bursitis, ganglionic cysts, strains, DeQuervain's disease, and carpal tunnel syndrome. Musculoskeletal disorders affecting the lumbar regions are also frequently reported in the workplace.

NIOSH has identified acute and chronic work-related musculoskeletal disorders as one of the ten leading occupational health problems affecting workers.[5] Accurate estimates of work-related CTD prevalence are not available because existing databases were not designed for the surveillance of occupational musculoskeletal disorders.[6] The available injury reporting databases "confound occupational and non-occupational related disorders, fail to differentiate between acute and chronic injuries, and lack standard terminology and diagnostic criteria for defining cumulative trauma-related musculoskeletal disorders."[6] Nonetheless, it was reported in 1983 that musculoskeletal injuries account for one-third of annual workers compensation claims, and the prevalence of job-related musculoskeletal disorders has increased through the 1980's.[6,7]

Studies have shown that work-related musculoskeletal disorders of the upper extremities can be associated with job activities that require: (1) repetitive movements of the upper extremities; (2) forceful grasping or pinching of tools or other objects by the hands; (3) awkward positions of the hand, wrist, forearm, elbow, upper arm, shoulder, neck, or head; (4) direct pressure over the skin and muscle tissue; and/or (5) use of vibrating hand-held tools.[8,9,10] Postures often associated with UE CTDs are extension, flexion, ulnar, and radial deviation of the wrist; open-hand pinching; twisting movements of the wrist and elbow; shoulder abduction; and reaching over shoulder height.

Health Hazard Evaluation Report No. 92-001

Activities associated with UE CTDs are frequently observed in many manufacturing and assembly jobs in industry. Occupations associated with a high incidence of CTDs include, but are not limited to, electronic components assembly, garment manufacturing, automobile assembly, small appliance manufacturing and assembly, and meat and poultry processing.[11,12,13] Because repetitive and forceful movements are required in many service and industrial occupations, new occupational groups at risk for developing UE CTDs continue to be identified. CTDs can also be associated with non-occupational activities or pre-existing conditions. For example, individuals with diabetes, rheumatoid arthritis, certain thyroid conditions, and kidney dysfunction appear to be at an increased risk for developing carpal tunnel syndrome.[14]

One of the most disabling CTDs is carpal tunnel syndrome (CTS). CTS was first reported as a clinical entity as early as 1854; however, it was not fully described in the medical literature until 1927.[15] CTS is caused by compression of the median nerve inside the carpal tunnel at the wrist. Clinical manifestations include pain, numbness, burning and/or tingling sensations in the hand and fingers in the distribution of the median nerve. At advanced stages, atrophy of the thenar muscle may occur.[16,17]

Previous investigations have focused on the role of occupational risk factors in CTD development. One study reported that workers performing jobs with force levels of 4 kilograms (kg) or more were four times as likely to develop hand/wrist CTDs as those workers with jobs requiring muscular exertions of 1 kg or less.[10] Job tasks with cycle times of 30 seconds or less were associated with a three times greater incidence of UE CTDs than jobs where cycle time was more than 30 seconds.[10] In addition, several cross-sectional and case control retrospective studies have examined the association between job risk factors and musculoskeletal morbidity.[18,19,20,21,22,23] In general, these studies support the association between physical risk factors (repetition, force, awkward postures, and vibration) and CTDs.

The complexity of repetitive motion patterns has made it difficult to establish threshold limits for defining CTD risk. Previous reports have used various definitions of repetitiveness to distinguish between different jobs.[5] These definitions are intended merely to assist in judging the relative risk of hand intensive jobs. Other risk factors, such as awkward or stressful posture and increased level of muscular force, can exacerbate the CTD risk.

Control measures to reduce the probability of developing these disorders should be directed at the physical risk factors. The preferred control method are engineering controls that reduce ergonomic stresses; however, administrative controls such as work enlargement, job rotation, etc., can be used as an interim measures. Surveillance of CTDs (including the use of health-care-provider reports) can aid in identifying high-risk workplaces, occupations, and industries and in directing appropriate preventive measures.[24]

Health Hazard Evaluation Report No. 92-001

Low Back Pain

Low back pain is one of the most common and most costly work-related musculoskeletal disorders in the United States.[5,25,26] Occupational risk factors for low back injuries include manual handling tasks,[27] twisting,[28] bending,[28] falling,[29] reaching,[30] lifting excessive weights,[28,31,32] prolonged sitting,[29] and vibration.[28,33] Some nonoccupational risk factors for low back injury include obesity,[34] genetic factors,[35] and job satisfaction.[36,37]

Control and prevention of job-related low back pain can be accomplished through the evaluation of job activities, identification of occupational risk factors, and implementation of appropriate interventions. Excessive bending, twisting or reaching, handling of excessive loads, prolonged sitting, and exposure to vibration are recognized risk factors for back injuries. Redesign of jobs can lead to the reduction of these risk factors, and good job design initially can prevent or minimize the occurrence of back injuries.[38] Workstation and job task design should consider: (1) the optimum work height (from waist to elbow height) to reduce bending and reaching; (2) logical spacial arrangement in conjunction with task sequence to reduce twisting; (3) provision of sit/stand stations to reduce prolonged postures; (4) appropriate package design and size to allow the worker to hold the lifted load close to the body with an effective coupling; and (5) package weight not exceeding typical human capabilities.[39] Multiple approaches such as job redesign, worker placement, and training may be the best methods for controlling back injuries and pain.[40]

VI. EVALUATION METHODS

The medical evaluation component of this HHE consisted of confidential health interviews with AT-6 workers, review of the Occupational Safety and Health Administration log of occupational injuries and illnesses (OSHA form 200), and informal discussions with the employees regarding musculoskeletal hazards associated with their jobs. The ergonomic assessment of the AT-6 press operation included measurement of workstation dimensions, a work methods analysis, and a biomechanical evaluation aided by videotape and still photography of the AT-6 process.

The videotape of the AT-6 press operation was analyzed by both regular speed and slow motion play-back. This work methods analysis was used to determine the work content of the job and to recognize occupational risk factors for CTD. The CTD stressors included in this evaluation were repetitive movements, muscular force, awkward and static posture, mechanical stress, and vibration.

Biomechanical evaluation was utilized to identify job tasks associated with an elevated risk of developing musculoskeletal disorders. A manual video digitizing technique was used to quantify work postures. The University of Michigan Strength Prediction Program was used to assess whether allowable physical limits for body segments were exceeded for a sub-group of the worker population.[41]

Health Hazard Evaluation Report No. 92-001

Body segments that were evaluated in the biomechanical evaluation included the upper extremities, back, hip, and lower extremities.

VII. RESULTS AND DISCUSSION

A. Medical Evaluation

1. Employee Profile

On the first day of the NIOSH site visit, ten employees were assigned to work on the AT-6 machine; all were interviewed. (On the second day of the NIOSH visit an 11th worker was assigned to the AT-6. Because this was the worker's first day on the AT-6, no attempt was made for inclusion in the interview process.)

The ten AT-6 employees had an average age of 29, with a range from 19 to 48. Eight of these workers were men, two were women. SANCAP operations started in February 1992, and eight AT-6 employees were hired at that time (including two former SIA America employees). One AT-6 employee was hired in June, and one at the beginning of September. All ten employees spent the majority of their time on the AT-6 machine, with occasional rotation to non-AT-6 tasks. Only two of the current employees had previous experience on the AT-6 machine under the former SIA America company.

Eight of the ten AT-6 employees had previously worked as press operators at one time or another. Three employees were on a regular six week rotational schedule: two weeks on one side of the press, two weeks on the other side of the press, and two weeks on a non-press task. Two other employees work 4-10 hours per week, providing lunch and relief breaks to the primary press operators. Two additional employees fill in as substitute press operators as needed. Another employee spent one month as a press operator, but was then assigned to other AT-6 tasks.

2. Confidential Interview Results

None of the ten interviewees reported sustaining chronic musculoskeletal injuries on the AT-6 machine. Two individuals reported occasional back difficulties but did not believe there was any association with work.

One interviewee was concerned with the potential of injury from the return of the carriage during operation of the press. The concern was that a striking injury could occur, with the carriage hitting an employee on its return path. Another employee knew of one unintentional injury occurring on the AT-6 machine under the previous facility owner; this involved an employee stepping on a cart in the hanging area.

Removable guard rails are available for use by the press operators. Certain belt sizes preclude the use of the rails due to potential damage from the product hitting the rail, otherwise rails may be installed and removed at the employee's discretion. Employees reported using the rails primarily when the belt size required the platforms to be further away from the press, thus necessitating leaning to reach the press. Rails were also used when the belts were heavy, either due to absolute belt size or due to the weight of the grit and backing material. Employees remarked that the rails were not always comfortable to work with, and some opted to never, or rarely, use the rails.

3. **OSHA 200 Log**

There were 12 entries in the OSHA 200 log between July 1, 1992, and September 22, 1992; none involved AT-6 employees. From February 1992 until May 1992, none of the AT-6 workers were SANCAP employees. They were employees of a temporary employment agency, and SANCAP did not include them in the OSHA 200 log during that time period. OSHA 200 logs for the period when SIA America operated the facility were not available.

B. **Ergonomic Evaluation**

1. **Work Methods Analysis**

The operators performed several steps to convert the pre-cut abrasive sheets to abrasive belts. According to management records, approximately 550 belts are produced each 10-hour shift. During the NIOSH site visit, the observed 10-hour shift of press operators included a one hour break and machine down-time for changing belt products. Motion analysis of the videotape indicates that the work cycle time for press operators during production periods was approximately 30 seconds per belt while manufacturing belts using sheets measuring 52" width x 103" length. The weight of each of these belts (52" width x 103" length) produced on September 22-23, 1992, was approximately 14 pounds. Table 1 describes the basic steps used at the AT-6 press to convert abrasive sheets to belts. The job task sequence for this process is diagrammed in Figure 1.

Press operation required frequent extended reaches, trunk twisting and bending, hand/wrist bending, and pinch grips. During the initial walk-through evaluation on the morning of September 22, the NIOSH researchers found that the AT-6 press operations were repetitive (30 seconds per cycle) and, at times, required awkward and static postures. The operators adopted prolonged awkward postures (approximately 50 degrees of trunk flexion) while positioning and matching the two ends of the abrasive sheets on the AT-6 press during

Health Hazard Evaluation Report No. 92-001

a time when the guard rail was used between the worker and the press. Awkward postures of the torso have been shown to be linked with increasing the likelihood of developing CTDs. Working with the torso bent forward, backward, or twisted can place excessive stress on the low back.[38,39] Repetitive tasks, extended reaches, pinch grips, and repeated hand/wrist bending are also recognized risk factors for contributing to the development of CTDs.[9]

Some operators frequently pounded the abrasive material with the palms of their hands (or clinched fist) a few times while constructing each joint of the abrasive belts. This impact action can result in mechanical stress on the hands. In addition, the work procedure required intense eye-hand coordination and concentration with sustained exertion force of the hands, including repeated and prolonged pinch grips. This may cause the workers to maintain a posture for extended periods of time resulting in static loading of the hand, wrist, neck, shoulder, and back.

During the videotaping sessions on September 22 and 23, the operators were encouraged to adjust the press platforms to accommodate more comfortable work postures. Five operators were videotaped for two hours operating the AT-6 press during a time when the guard rails were not employed. Only one operator worked on the left side during the time period of videotaping. The other four operators worked sequentially on the right side while videotaping.

Two operators were selected for videotape posture analyses: the only operator who worked on the left side, and the operator on the right side who was videotaped for the longest period of time. Each measured observation was made on a separate section of the videotape. Without the side rails installed, trunk flexion angles were measured to be an average of 32 ($\pm$ 6) degrees based on six observations of the left side operator. An average of 25 ($\pm$ 8) degrees was found for the right side operator based on six observations.

After several adjustments of the platforms, the worker on the left side performed the job with an average of 19 ($\pm$ 5) degrees of trunk flexion based on 12 observations. The worker observed on the right side attained an average of 14 ($\pm$ 4) degrees of trunk flexion based on six observations. These reductions in torso flexion demonstrate the importance of properly adjusted platforms to accommodate the individual worker to the job conditions.

The two operators were also requested to install the side rails on the platforms and adjust the rail height to accommodate comfortable work postures. Based on six observations of each operator with the side rails installed, the workers were able to perform the work with an average of 25 ($\pm$ 6) and 19 ($\pm$ 3) degrees of trunk flexion on the left and right sides, respectively. The large difference between the measured flexion of the torso with the posture approximated during the initial walk-through provides an excellent example of the stress imposed by an improperly adjusted guard rail and work platform.

2. **Biomechanical Evaluation**

The biomechanical evaluation using the University of Michigan 2D Static Strength Prediction Program is presented in Figure 2. This analysis showed that the compressive forces for the low back and the strength required to perform the task did not exceed back compression design limits of 770 lbs. However, analyses of the postures assumed in operating the press indicated that the values for the hip, knee, and ankle for some postures exceeded the Strength Design Limit (SDL) reported in the 2D Static Strength Prediction Model published by the University of Michigan. This analysis revealed that since some work position postures exceeded the SDL, greater than 1% of men and 25% of women will not be able to attain the necessary strength to operate the AT-6 press with the observed postures. In addition, the biomechanical assessment revealed that some of the postures and force parameters examined would cause the workers to fall forward (i.e., postural imbalance).

C. **Safety Issues**

1. **Falling Hazard**

There is an open space between the two platforms at the AT-6 press, creating a potential hazard of falling from the platforms to the shop floor. The distance from the platforms is approximately 3-3½ feet above the floor. A previous owner of the facility developed the side rails to prevent falls. Press operators believed the side rails were present to support body weight, and to reduce the development of fatigue. Shorter stature operators stated that the side rails restricted their performance especially during the manufacture of large size belts. Side-rails were reported to be rarely used. The video analysis results indicate that the operators can work in more comfortable postures if they adjust the platforms (height and horizontal distance to the machine), as well as the side rails (height).

2. **Tripping/Falling Hazard: Accommodation for Disability**

The AT-6 machine has several sets of fixed industrial steps leading from the floor up to the work platform. There are also two ladders with rungs, fixed to either side of the movable platforms located at the AT-6 press. One employee was noted to have a below-the-knee amputation with a fitted prosthesis. This same employee was working as a relief press operator during the NIOSH site visit, and was observed ascending and descending the ladders. Ladder use is a potential safety

Health Hazard Evaluation Report No. 92-001

hazard for this individual due to the decreased maneuverability and control of a prosthetic leg. In this circumstance, ascending and descending via the stairs would decrease the likelihood of unintentional injury, and would be an appropriate accommodation for this employee's disability.

VIII. RECOMMENDATIONS

A. Engineering Controls

1. The side rails should be in place at the AT-6 press at all times to prevent falls from the elevated platforms to the shop floor.

2. Currently the side rail height can be adjusted from 25" to 46". The lower height limit needs to be decreased to 23" in order to accommodate workers of shorter stature. The hardware of the rails should be redesigned to allow workers to make the adjustments with ease. Also, padding on the rail should be installed to reduce the mechanical pressure when a worker leans on the rail. Improving ease of adjustability and reducing discomfort associated with the use of the rail should encourage workers to utilize the guard rail properly, thereby reducing the static load on the lumbar region.

3. The use of bare hands to apply blunt mechanical force on the belt joint should be prohibited. Use of a rubber hammer or applying pressure to the joint of the belts by hand, without impact, to apply force is recommended. Repeated pounding with the palm of the hand is a risk factor for contributing to the development of CTD of the hand and wrist.

4. Install an inner "product" deflector bar, affixed to the press portion of the AT-6, to keep the abrasive belts toward the center line of the machine during the joint pressing process. This will allow the operators to move the platforms closer to the press table, thus reducing awkward trunk flexion. The deflector bar can be constructed of rubber, with rollers, or otherwise be designed to protect the product from contacting the platform and becoming damaged.

5. Provide a stool or install adjustable sit/lean workstations for the press operators so that they can relax fatigued muscle groups during portions of the work cycle.

6. A foot rail can also be installed on the work platforms below the AT-6 press so that operators rest one foot and change body posture to reduce the static loading of the lower extremities and back. The rail should be 4-6" above the platform surface.

Health Hazard Evaluation Report No. 92-001

B. **Administrative Controls**

1. Employees should be encouraged to adjust the side rails and platform to comfortable positions. Evaluation and recording of the platform and side rail positions should be established. The information to be recorded should include:

   a. identification of operator;

   b. platform position;

   c. rail height;

   d. length and width of product;

   e. grit type and backing of product.

   An individualized index of comfortable settings for each press operator, for multiple sizes and types of belts, may then be generated. Use of the information should facilitate making adjustments quickly for different production runs. Affixing measurement lines (or tape) to the floor, platform, and rails may be useful, both while developing the initial record, and to facilitate quick adjustments later.

2. All employees working as press operators should receive training on ergonomic hazards. Training should include the use of side rails and platforms at the AT-6 press. Workers should be able to demonstrate proficiency in making platform and side rail adjustments on-the-job. The use of the individualized indexes, as discussed in the paragraph above, should also be covered. New employees, or employees transferring to the AT-6 press, should receive similar training prior to beginning press work. An individualized index of settings should be developed for each new press operator.

3. Establish a medical surveillance program to evaluate workers' health status in regards to CTDs, including back injuries. Early intervention CTDs often results in more successful treatment of the injured worker.

C. **Work Practices**

1. Press operators and relief operators should adjust the platforms and side rails to obtain a comfortable posture before beginning operation. They should re-adjust the positions as needed during the working period.

2. Operators should avoid overreaching and excessive torso twisting while picking up the pre-cut abrasive sheets from the conveyor. Walking one step further to the transfer conveyor to obtain the abrasive

sheet will reduce this problem. Overreaching may result in excess musculoskeletal stress and possibly injury.

3. Employees with disabilities which make ladder climbing a potential safety hazard should ascend and descend to the work platforms via the stairs which are available.

IX. REFERENCES

1. Conn HR [1931]. Tenosynovitis. *Ohio State Medical Journal* 27:713-716.

2. Pozner H [1942]. A report on a series of cases on simple acute tenosynovitis. *Journal of Royal Army Medical Corps* 78:142.

3. Hymovich L, Lindholm M [1966]. Hand, wrist, and forearm injuries. *Journal of Occupational Medicine* 8:575-577.

4. NIOSH [1977]. Cincinnati, OH: U.S. Department of Health and Human Services, Public Health Service, Centers for Disease Control, National Institute for Occupational Safety and Health, Division of Surveillance, Hazard Evaluations, and Field Studies, HETA Report No. TA 76-93.

5. DHHS [1983]. Prevention of leading work-related diseases and injuries - musculoskeletal injuries. Atlanta, GA: U.S. Department of Health and Human Services, Public Health Service, Centers for Disease Control, National Institute for Occupational Safety and Health, Division of Surveillance, Hazard Evaluations, and Field Studies. *Morbidity and Mortality Weekly Report* Vol. 32, No. 14, April 15.

6. DHHS [1989]. Proposed national strategies for the prevention of leading work-related diseases and injuries: musculoskeletal injuries. Cincinnati, OH: U.S. Department of Health and Human Services, Public Health Service, Centers for Disease Control, National Institute for Occupational Safety and Health, DHHS (NIOSH) Publication No. 89-129.

7. DHHS [1992]. 1991 conference summary: a national strategy for occupational musculoskeletal injuries: implementation and research needs. Cincinnati, OH: U.S. Department of Health and Human Services, Public Health Services, Center for Disease Control, National Institute for Occupational Safety and Health, DHHS (NIOSH) Publication No. 93-101, November.

8. Armstrong TJ [1986]. Ergonomics and cumulative trauma disorders. *Hand Clinics* 2:553-565.

9. Putz-Anderson V (Editor) [1988]. Cumulative trauma disorders: a manual for musculoskeletal diseases of the upper limbs. New York, NY: Taylor & Francis.

10. Silverstein BA, Fine LJ, Armstrong TJ [1987]. Hand-wrist cumulative trauma disorders in industry. *British Journal of Industrial Medicine* 43:779-784.

11. Armstrong T, Foulke J, Joseph B, Goldstein S [1982]. Investigation of cumulative trauma disorders in a poultry processing plant. *American Industrial Hygiene Association Journal* 43:103-116.

12. NIOSH [1989]. John Morrell & Co., Sioux Falls, SD. Cincinnati, OH: U.S. Department of Health and Human Services, Public Health Service, Centers for Disease Control, National Institute for Occupational Safety and Health, Division of Surveillance, Hazard Evaluations, and Field Studies, HETA Report 88-180-1958.

13. Habes DJ, Putz-Anderson V [1985]. The NIOSH program for evaluating hazards in the workplace. *Journal of Safety Research* 16:49-60.

14. Cannon LE, Bernacki E, Walter S [1981]. Personal and occupational factors associated with carpal tunnel syndrome. *Journal of Occupational Medicine* 23:255-258.

15. Pfeffer GB, Gelberman RH, Boyes JH, Rydevik B [1988]. The history of carpal tunnel syndrome. *Journal of Hand Surgery* 13B:28-34.

16. Phalen GS [1972]. The carpal tunnel syndrome. *Clinical Orthopaedics* 83:29-40.

17. Spinner RJ, Bachman JW, Amadio PC [1989]. The many faces of carpal tunnel syndrome. *Mayo Clinic Proceedings* 64:829-836.

18. Anderson JA [1972]. System of job analysis for use in studying rheumatic complaints in industrial workers. *Annals of Rheumatologic Diseases* 31:226.

19. Hadler N [1978]. Hand structure and function in an industrial setting. *Arthritis and Rheumatism* 21:210-220.

20. Drury CD, Wich J [1984]. Ergonomic applications in the shoe industry. In: *Proceedings of the International Conference on Occupational Ergonomics*, Toronto, May 7-9, pp. 489-493.

21. Cannon L [1981]. Personal and occupational factors associated with carpal tunnel syndrome. *Journal of Occupational Medicine* 23(4):225-258.

22. Armstrong TJ, Foulke JA, Bradley JS, Goldstein SA [1982]. Investigation of cumulative trauma disorders in a poultry processing plant. *American Industrial Hygiene Association Journal* 43:103-106.

23. Silverstein BA [1985]. The prevalence of upper extremity cumulative trauma disorders in industry. Ph.D. Dissertation, University of Michigan.

24. Cummings J, Maizlish N, Rudolph MD, Dervin K, Ervin [1989]. Occupational disease surveillance: carpal tunnel syndrome. *Morbidity and Mortality Weekly Report*, July 21, pp. 485-489.

25. Holbrook TL, Grazier K, Kelsey JL, Stauffer RN [1984]. The frequency of occurrence, impact, and cost of selected musculoskeletal conditions in the United States. Chicago, IL: American Academy of Orthopaedic Surgeons.

26. Snook SH [1983]. Back and other musculoskeletal disorders. Chapter 23 in: Occupational Health, Levy BS, and Wegman DH (editors). Boston, MA: Little, Brown and Company.

27. Bigos SJ, Spenger DM, Martin NA, Zeh J, Fisher L, Machemson A, Wang MH [1986]. Back injuries in industry: a retrospective study. II. Injury Factors. *Spine* 11:246-251.

28. Frymoyer JW, Cats-Baril W [1987]. Predictors of low back pain disability. Clinical Orthopaedics and Rel Research 221:89-98.

29. Magora A [1972]. Investigation of the relation between low back pain and occupation. *Industrial Medicine and Surgery* 41:5-9.

30. BLS [1982]. Back injuries associated with lifting. Washington, DC: U.S. Department of Labor, U.S. Government Printing Office, Bureau of Labor Statistics Bulletin 2144.

31. Chaffin DB, Park KS [1973]. A longitudinal study of low-back pain as associated with occupational weight lifting factors. *American Industrial Hygiene Association Journal* 34:513-525.

32. Liles DH, Dievanyagam S, Ayoub MM, Mahajan P [1984]. A job severity index for the evaluation and control of lifting injury. *Human Factors* 26:683-693.

33. Burton AK, Sandover J [1987]. Back pain in grand prix drivers: a found experiment. *Ergonomics* 18:3-8.

34. Deyo RA, Bass JE [1989]. Lifestyle and low-back pain: the influence of smoking and obesity. *Spine* 14:501-506.

35. Postacchini F, Lami R, Publiese O [1988]. Familial predisposition to discogenic low-back pain. *Spine* 13:1403-1406.

36. Bureau of National Affairs, Inc. [1988]. *Occupational Safety and Health Reporter* July 13, pp. 516-517.

37. Svensson H, Andersson GBJ [1989]. The relationship of low-back pain, work history, work environment, and stress. *Spine* 14:517-522.

38. NIOSH [1990]. Harley-Davidson, Inc., Milwaukee, WI; Cincinnati, OH: U.S. Department of Health and Human Services, Public Health Service, Centers for Disease Control, National Institute for Occupational Safety and Health, Division of Surveillance, Hazard Evaluations, and Field Studies, HETA Report 90-134-2064.

39. NIOSH [1981]. Work practices guide for manual lifting. Cincinnati, OH: U.S. Department of Health and Human Services, Public Health Service, Centers for Disease Control, National Institute for Occupational Safety and Health, DHHS (NIOSH) Publication No. 81-122.

40. Snook SH [1987]. Approaches to the control of back pain in industry: job design, job placement, and education/training. *Spine: State of the Art Reviews* 2:45-59.

41. The University of Michigan [1990]. 2D static strength prediction program™. The Center for Ergonomics, The University of Michigan, Ann Arbor, MI 48109.

X. AUTHORSHIP AND ACKNOWLEDGEMENTS

Report prepared by:
Faye T. Bresler, M.D., M.P.H.
Medical Officer
Medical Section
Hazard Evaluations and Technical
  Assistance Branch
Division of Surveillance, Hazard
  Evaluations and Field Studies

Hongwei Hsiao, Ph.D.
Research Biomechanical Engineer
Protective Equipment Section
Protective Technology Branch
Division of Safety Research

Health Hazard Evaluation Report No. 92-001

Kevin Hanley, M.S.P.H., C.I.H.
Industrial Hygienist
Industrial Hygiene Section
Hazard Evaluations and Technical
  Assistance Branch
Division of Surveillance, Hazard
  Evaluations and Field Studies

Originating Office: Hazard Evaluations and Technical
  Assistance Branch
Division of Surveillance, Hazard
  Evaluations and Field Studies

XI. DISTRIBUTION AND AVAILABILITY OF REPORT

Copies of this report may be freely reproduced and are not copyrighted. Single copies of this report will be available for a period of 90 days from the date of this report from the NIOSH Publications Office, 4676 Columbia Parkway, Cincinnati, Ohio 45226. To expedite your request, include a self-addressed mailing label along with your written request. After this time, copies may be purchased from the National Technical Information Service (NTIS), 5285 Port Royal, Springfield, Virginia 22161. Information regarding the NTIS stock number may be obtained from the NIOSH Publications Office at the Cincinnati address.

Copies of this report have been sent to:

1. SANCAP Inc., Alliance, Ohio
2. International Chemical Workers Union, International, Akron, Ohio
3. OSHA, Region V

For the purpose of informing affected employees, copies of this report shall be posted by the employer in a prominent place accessible to the employees for a period of 30 calendar days.

Table 1
Description of Motions Used by AT-6 Press Operators
SANCAP ABRASIVES, Inc.
HETA 92-001

Step	Number of repetitions	Description	Body part	Action
1	1	Get pre-cut abrasive sheet from AT-6 conveyor	right hand	pinch grip
2	1	Carry pre-cut abrasive strip to AT-6 press	both hands	pinch grip
3	1	Position and match the two ends of the abrasive strip on AT-6 press to form belt	both hands	pinch grip
4	1 or multiple	Apply pressure to the joint of the abrasive belt	both hands	finger or palmar pressure or pounding
5	1	Side-step on switches to activate press	one foot	step
6	1	Use brush to remove loose abrasive material	right hand	pinch grip
7	1	Remove any remaining adhesive from the press table or abrasive belt	left hand	pinch grip

**Figure 1
Work Sequence at AT-6 Press
SANCAP ABRASIVES, Inc.
HETA 92-001**

Step 1

Step 2

Step 3

Step 4

**Figure 1
(continued)
Work Sequence at AT-6 Press
SANCAP ABRASIVES, Inc.
HETA 92-001**

Step 5

Step 6

Step 7

Repeat Step 1

Figure 2
Biomechanical Evaluation
University of Michigan, 2D Static Strength Prediction Program
SANCAP ABRASIVES, Inc.
HETA 92-001

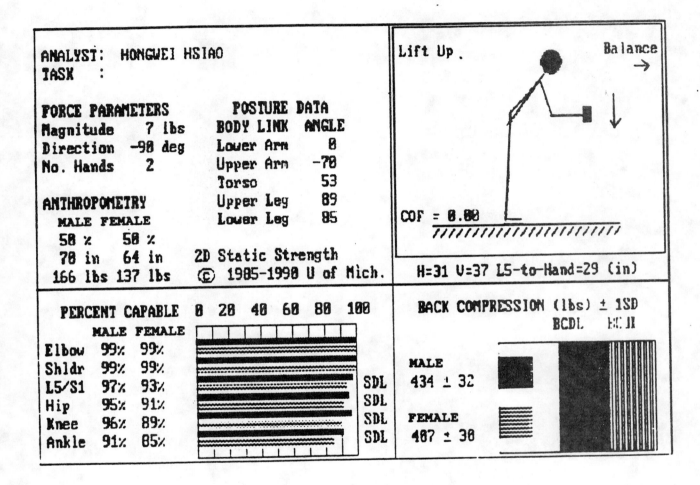

HETA 93-0805-2387
FEBRUARY 1994
UNICCO
HARTFORD, CONNECTICUT

NIOSH INVESTIGATORS:
Thomas J. Van Gilder, M.D.
Thomas Waters, Ph.D.
Peter Fatone, M.S.
Sherry Baron, M.D., M.P.H.

SUMMARY

The National Institute for Occupational Safety and Health (NIOSH) received a request for a Health Hazard Evaluation (HHE) on March 22, 1993, from the Service Employees International Union (SEIU) regarding the use of a back-pack vacuum cleaner (BPVC) among employees of the UNICCO Service Company at the Travelers' Insurance complex in Hartford, Connecticut.

SEIU members have expressed concerns about BPVC use in a number of locations throughout the country, including the Travelers' Insurance complex, a series of office buildings consisting of standard, modern office space. These concerns have centered around perceived health effects of BPVC use, including musculoskeletal (primarily shoulder and back) problems, noise, heat, and vibration.

On May 12, 1993, the NIOSH investigator conducted a walk-through inspection of typical work areas in the Travelers' Insurance complex, viewed and videotaped a demonstration of BPVC use by two UNICCO employees in simulated offices, and interviewed three UNICCO employees concerning work practices and symptoms. Laboratory analysis of the BPVC, including a biomechanical assessment, and measurements of vibration, heat, and noise, was performed by NIOSH personnel. The three workers interviewed reported similar problems with the BPVC. These complaints consisted primarily of musculoskeletal (especially shoulder and back) discomfort and the sensation of excessive heat associated with the BPVC. Laboratory analysis revealed increased biomechanical stress and heat in a person wearing the BPVC compared to a person not wearing the BPVC. Noise and vibration were within recommended exposure limits. We were unable to determine the health risk posed by the combined effects of the various stressors.

NIOSH investigators found a number of complaints associated with the use of back-pack vacuum cleaners. Laboratory analysis of the vacuum cleaner revealed biomechanical stress and heat exposure; noise and vibration were within recommended exposure limits. We were unable, however, to determine the health risk posed by the combined effects of the various stressors.

KEYWORDS: SIC 7349 (janitorial services), back-pack vacuum cleaners, ergonomics, biomechanical stress, vibration, heat, noise

Hazard Evaluation and Technical Assistance Report No. 93-0805

INTRODUCTION

The National Institute for Occupational Safety and Health (NIOSH) received a request for a Health Hazard Evaluation (HHE) on March 22, 1993, from the Service Employees International Union (SEIU) regarding the use of a back-pack vacuum cleaner (BPVC) among employees of the UNICCO Service Company working at the Travelers' Insurance complex in Hartford, Connecticut. On May 12, 1993, NIOSH conducted a site visit to meet with SEIU, UNICCO, and BPVC manufacturer representatives, and to view and videotape the use of the BPVC. Subsequently, a BPVC (Pro-Team Inc.'s Quarter-Vac™) was analyzed by NIOSH personnel in Cincinnati for biomechanical stress, vibration, heat, and noise incident to its use. This report provides the final results of our assessment of the BPVC in use by UNICCO employees at the Travelers' Insurance complex.

BACKGROUND

Facility Description
The Travelers' Insurance Corporate complex is a typical modern high-rise office building. The interior space is divided into individual offices by temporary dividers. There are formal conference rooms on most floors. Standard modern office equipment (e.g., desks, personal computers, copiers) is located on each floor.

Process Description
Employees of contract cleaning companies typically begin their workday after the offices are closed and work for approximately eight hours. The workers are responsible for a variety of tasks, including vacuuming the floors, emptying waste baskets, and removing trash. Over the past several years, the BPVC has become increasingly popular among contract cleaning companies, expanding from its original use in confined spaces such as aircraft. The BPVC typically is used 8 hours per night, 4 days per week. It is used to clean especially dirty areas throughout the office building. Workers vacuum under desks and tables, which requires periods of forward bending, as shown in figure 6. They also vacuum stretches of open hallway; this is performed in the upright position. The BPVC is plugged into a wall socket via a long extension cord which the workers bend to plug, unplug, and coil several times per shift. Workers report having two scheduled breaks during which they remove the vacuum cleaner; they also report removing the vacuum cleaner an average of once per hour (for several minutes) to rest their backs and to get a break from the heat. Reportedly, the vacuum does not become appreciably heavier as it fills with dirt; workers empty the unit an average of twice per night.

Hazard Evaluation and Technical Assistance Report No. 93-0805

EVALUATION PROCEDURES

Ergonomic

Biomechanical Assessment
Biomechanical evaluation provides a method for predicting the magnitude of muscle, ligament, and joint forces developed within the body as a result of external loads or gravitational forces.

A biomechanical analysis was performed to identify and quantify the physical loading of the spine due to the use of a BPVC. Figure 1 shows the BPVC with approximate dimensions. To simplify the evaluation, a static, two-dimensional analysis of an average male worker was used to estimate the amount of additional loading imposed on the spine, back, and abdominal and back muscles as a result of wearing the vacuum cleaner. The biomechanical assessment evaluated two different postures, upright standing and 85 degrees of forward flexion, which were postures observed during the site visit. The cleaner weighed 15 pounds when the bag was full with typical dust and dirt. For computational purposes, the worker height and weight were assumed to be 68.7 inches and 165 pounds (i.e., average US male). Input for the biomechanical analysis consisted of estimates of body geometry and weight distribution and of BPVC geometry and weight distribution relative to the worker.

The first step of the analysis consisted of isolating the upper portion of the body with a horizontal plane passing through the L5/S1 intervertebral joint (i.e., joint between the fifth lumbar and first sacral vertebral bodies). A sketch of this isolated section is shown in figure 2, where W_T represents weight of the upper torso, head, and arms at the center of gravity of this body section; W_V represents the weight of the BPVC at the center of gravity of the cleaner; F_B represents the back extensor muscle force; F_A represents the abdominal flexor muscle force; and a triangle ($\triangle$) represents the inter-vertebral joint between the fifth lumbar (L5) vertebral segment and first sacral vertebral segment (S1) (labeled L5/S1). The triangle symbolizes a fulcrum for a simple lever system. The forces W_T and W_V created moments (i.e., the rotational tendency caused by application of a perpendicular force on a lever arm at some distance from its axis of rotation) about the fulcrum. These moments were balanced by the muscle forces, F_A and F_B, and equilibrium was achieved. All of these forces resulted in an upward reaction force (R) at the fulcrum. The reaction force, R, represents the axial compression force on the L5/S1 intervertebral disc. (If the muscle force or compressive force is too high, the musculoskeletal tissues may be damaged and result in back discomfort, pain, or even injury.)

Hazard Evaluation and Technical Assistance Report No. 93-0805

To simplify the analysis, free-body diagrams were developed and are shown in figures 3-5, 7 and 8. Free-body diagrams enhance the visualization of the biomechanical effects of postural changes and external loads. Based on these diagrams, mechanical equations were developed and solved to find the unknown muscle forces (F_B and F_A) and the reaction force (R). According to the laws of static equilibrium, the sum of the moments and forces must be equal to zero. Since the weights and distances were known, the muscle forces and compressive force at the fulcrum needed to maintain static equilibrium (i.e., no body movement) were calculated for each posture.

Vibration

We used an accelerometer to measure vibration in the back pack. No standards exist for local (i.e., non-whole-body) vibration. The subjects were assumed to be exposed to whole-body vibration, although not in the purest form. This was considered a reasonable conjecture since a large portion of the back was exposed to the vibration and the conditions more closely resembled whole-body vibration than segmental vibration.

In order to evaluate the potential adverse health effects due to BPVC vibration exposure, the magnitude of vibrational accelerations was measured across a spectrum of frequencies for a worker wearing the BPVC. Two types of measurements were made. The first type collected vibration data on the BPVC near the location where the device contacts the lower back. The second type measured the amount of vibration at the interface of the worker and the BPVC.

For the first set of measurements, three accelerometers were attached in an orthogonal arrangement to the support bushing near the base of the vacuum cleaner. These accelerometers collected data in all three directions. The coordinate system and accelerometer placement are shown in figure 9. Two subjects were tested in three different postures. The subjects stood erect with back at 0 (v), 45 (f), and 90 (n) degrees to the Z axis.

For the second set of measurements, a flat, rubberized, tri-axial accelerometer was positioned directly between the BPVC's hard, plastic harness plate and the subject's back. The accelerometer measured the amplitude and frequency of the vibration that was transferred between the worker and the vacuum. For this evaluation, one subject was tested in three different pos-tures, standing erect with the back at 0, flexed forward at 45 degrees, and flexed forward at 90 degrees to the Z axis. These postures simulated the full range of motion during typical working conditions.

Thermal

To assess the potential effects of heat buildup from the BPVC, an analysis was performed to determine if the BPVC significantly increased the temperature of the skin of a worker wearing the BPVC. This analysis consisted of attaching a thin, thermal skin sensor to the subject's back beneath the BPVC to record skin temperature. The experiment consisted of recording the temperature of the skin every 5 minutes for a 1-hour period after the BPVC was turned on. The measurements were made with a Yellow Springs Tele-thermometer (Model 43TA) with the subject in a quiet, sitting posture. The room temperature was approximately 73°F.

Noise
Sound measurements were performed on the BPVC to evaluate possible noise hazards to workers during operation of the BPVC. The test consisted of placing the BPVC on a mannequin in a sound-proof room and attaching a microphone to the right ear of the mannequin. Measurements were made for three conditions: (1) motor off (i.e., background); (2) motor on with the suction tube open; and (3) motor on with the suction tube blocked.

Medical
The NIOSH investigator watched and videotaped two employees using the BPVC in simulated offices. Three employees were interviewed; questionnaires designed and distributed by SEIU and returned by eight employees were reviewed.

RESULTS AND DISCUSSION

Ergonomic

Biomechanical
The free-body diagram shown in figure 3 describes the loading condition for an average male worker in an upright posture not wearing the BPVC. In this case, the weight of the trunk (W_T) was balanced over the fulcrum and the moment arm ($l1$) was approximately zero, which resulted in no moment. Thus, the flexor and extensor muscle and reaction forces were minimized, such that F_A and F_B were approximately equal to zero, and R was equal to W_T or 59 pounds.

Compare the loading condition in figure 3 (upright posture without BPVC) to that shown in figure 4 (upright posture with BPVC). In the case depicted in figure 4, the weight of the BPVC (W_V) was located behind the fulcrum at a distance of 13 inches, which created a positive moment that was balanced by an opposing negative moment created by the abdominal flexor muscle force, F_A. Assuming that both the BPVC moment arm ($l3$) and the abdominal muscle moment arm ($l2$) were 7 inches, the abdominal muscle force for static equilibrium was calculated to be 15 pounds. The resultant loading force (R) for this condition was equal to the sum of the applied forces $W_T + W_V + F_A$ (59 + 15 + 15), or 89 pounds. This represented a 50% increase in the compression force due to the BPVC.

Hazard Evaluation and Technical Assistance Report No. 93-0805

We observed that the loading condition described in figure 4 will generally not occur because a person typically will lean forward slightly when something heavy is placed on the back. This slight forward lean is used to balance the load and minimize the internal forces. This is clearly shown in figure 5, which depicts a worker flexed forward slightly with the BPVC. In order to balance the positive moment created by the BPVC, the worker flexes forward about 6.5 degrees to utilize the negative moment created by the weight of the trunk (W_T). In this posture, the muscle forces and the reaction force were reduced to 0 and 74 pounds (the sum $W_T + W_V$), respectively. In this posture, the resulting increase in compression force due to the BPVC was only about 25% greater than without the BPVC. This posture would be preferred by the worker because less energy would be expended due to the limited muscle force requirements and the disc-compression force is minimized. Although the muscle and reaction forces were minimized by this continuous forward flexion, other biomechanical factors, such as ligament strain or joint misalignment, may be increased. These factors may not be readily perceived by the worker and may increase a worker's risk of musculoskeletal injury. For example, in other industries, repetitive forward bending has been shown to increase the risk of low-back pain for workers.[1]

The worker shown in figure 6 illustrates a posture of extreme forward flexion often used while wearing the BPVC. Figures 7 and 8 describe the loading conditions for two cases of extreme forward flexion. Figure 7 depicts an 85 degree forward-flexion posture while not wearing the BPVC, and figure 8 depicts an 85 degree forward-flexion posture while wearing the BPVC. For the condition in figure 7, the estimated extensor muscle force was 269 pounds and the disc compression force was 294 pounds. When the BPVC was added, the estimated extensor muscle force (F_B) was increased to 330 pounds. The compressive component of the reaction force was increased to 336 pounds, which represented a 23% increase in muscle force and a 14% increase in disc-compression force. While the absolute magnitudes of the muscle forces and disc-compression force values did not necessarily indicate a high risk of low back injury, the relative increases in their magnitudes clearly demonstrated that muscular loading was increased, which could result in fatigued or strained muscles. Moreover, regardless of the magnitude of the reaction forces, increased repetitive loading to the joint surfaces due to the BPVC may increase the risk of a cumulative injury to the spine.

A summary of the results of the biomechanical analysis is presented in table 1.

Vibration
Results are shown in tables 2 and 3. Exposure to whole-body vibration has been associated with such health effects as headaches, fatigue, lack of concentration, and gastrointestinal disturbances.[2] In addition, exposure to whole-body vibration over a number of years may result in low-back pain.[3]

The sensitivity of the human body to the whole-body vibration depends on the frequency and direction of the exposure. To determine the potential health risks associated with these vibration levels, the actual data must be frequency weighted using weighting filter curves to approximate the effective acceleration levels. The International Standards Organization (ISO) has devised frequency weighting curves for both lateral and longitudinal whole-body vibration.[4,5] Refer to figure 10 for the curves used in this study. After adjusting the results of table 3 according to the ISO curves, the peak vector sum of the acceleration components for the ranges of 0.7-5 Hz and 5-20 Hz are 0.04 and 0.12 m/sec^2, respectively. According to the ISO Dose System for Whole Body Vibration, the "fatigue-decreased" exposure limit for the BPVC vibration is determined to be 24 hours.[5] It should be noted, however, that the "reduced comfort" exposure limit is less than 24 hours. Analysis of these results seem to suggest that vibration alone does not represent a health risk, but could result in discomfort to the worker.

Thermal
The results of the evaluation are displayed graphically in figure 11. The starting temperature was 93.5°F and the final temperature was 95.0°F. Although there was only a slight increase in temperature during the 1-hour period, when the subject was asked about his perception of the temperature, he indicated that the BPVC "felt warm."

Noise
None of the sound measurements exceeded 80 db, which is considered safe for exposures up to 32 hours.[6]

Medical
The three workers interviewed reported musculoskeletal discomfort in the neck, shoulder, upper and lower back which they related to the use of the BPVC. Two of the three workers interviewed stated that the BPVC caused their mid-backs to become hot and to perspire heavily. One of the workers stated that the vibration of the BPVC was uncomfortable, especially in the shoulders and lower back.

The workers surveyed by SEIU reported headaches, earaches, back pain and strain, joint pain, and "kidney problems."

The workers seemed to be uncertain about the proper use of the vacuum; the video taken showed that the workers were unaware of the proper wearing and adjustment of the unit. Many reported receiving little or no training in its use.

Hazard Evaluation and Technical Assistance Report No. 93-0805

CONCLUSIONS

The workers had a number of complaints related to the BPVC, including musculoskeletal pain and discomfort, excessive heat, excessive vibration, and excessive noise. These complaints were expressed by the workers interviewed by NIOSH as well as by those polled by SEIU. Laboratory evaluation of the BPVC revealed potential physiologic bases for these complaints. For example, depending on the posture of the worker wearing the BPVC, the BPVC may cause a 14%-50% increase in the vertebral-disc compression force, compared to that in a worker in the same position without the BPVC. Additionally, in order to compensate for the added weight of the BPVC, workers must adopt an awkward posture while wearing the BPVC. Despite this, the NIOSH investigators could identify no health hazard based on any of the factors observed (posture, motion) or measured (weight, heat, vibration, noise). That is, the measured exposures did not exceed any published standard nor is it apparent that BPVC use would lead to such health outcomes as disc herniation, nerve damage, or hearing loss. However, these factors (particularly biomechanical stress, vibration, and heat) may interact to create a situation more hazardous than any individual factor acting alone. In addition, these stressors may contribute to increased job dissatisfaction even in the absence of a clear health hazard.

The introduction of new equipment or new technologies into the workplace should create a healthier environment for the workers or at least create no new hazards. At this point, it is unclear whether the BPVC represents a hazard. However, given the number and type of worker complaints related to the BPVC, careful evaluation, by the workers and management, of this equipment and the reasons for its introduction into a given workplace is warranted. In those workplaces in which the BPVC is in use or being considered, following these two recommendations may reduce the number of complaints associated with its use:

a.) Train the workers in the proper use of BPVC. Periodically monitor its use and fit, as well as workers' health complaints and comfort, to ensure continued proper use.

b.) Allow the workers some flexibility in choosing the appropriate equipment for the task. One worker reported a number of difficulties using the BPVC in a confined space. These difficulties resolved when an upright unit was used.

REFERENCES

1. Snook SH, Campanelli RA, Hart JW. A study of three preventive approaches to low-back injury. Journal of Occupational Medicine. 1978;20(7):478-481.

2. Taylor W, Wasserman DE. Occupational vibration. Chapter 21. In: Zenz C, ed. Occupational Medicine Principals and Practical Applications. 2nd ed. Chicago: Year Book Medical Publishers, Inc, p. 326, 1988.

3. Hulshof C. Whole-body vibration and low-back pain. A review of epidemiologic studies. Int Arch Occup Environ Health. 1987;59:205-220.

4. International Standards Organization Draft Proposal 5349. Guide for the measurement and evaluation of human exposure to vibration transmitted to the hand, 1980.

5. International Standards Organization ISO 2631-1978. Guide for the evaluation of human exposure to whole-body vibration, 1978.

6. Berger EH, Ward WD, Morrill JC, and Royster LH eds. Noise & Hearing Conservation Manual. American Industrial Hygiene Association, pp. 546-547, 1988.

AUTHORSHIP AND ACKNOWLEDGMENTS

Evaluation Conducted and Report Prepared By:

Thomas J. Van Gilder, M.D.
Medical Officer
Medical Section
Hazard Evaluations and Technical
 Assistance Branch

Thomas Waters, Ph.D.
Research Physiologist
Applied Psychology and
 Ergonomics Branch
Division of Biomedical and
 Behavioral Science

Peter Fatone, M.S.
Mechanical Engineer
Applied Psychology and
 Ergonomics Branch
Division of Biomedical and
 Behavioral Science

Hazard Evaluation and Technical Assistance Report No. 93-0805

                                                            Sherry Baron, M.D.
Medical Officer
Medical Section
Hazard Evaluations and Technical Assistance Branch

Originating Office: Hazard Evaluations and Technical Assistance Branch
Division of Surveillance, Hazard Evaluations, and Field Studies

Laboratory Support: Curt Sizemore and Staff
Division of Biomedical and Behavioral Science

                                                          Tom Doyle
State of Ohio Bureau of Workman's Compensation
Columbus, Ohio.

REPORT DISTRIBUTION AND AVAILABILITY

Copies of this report may be freely reproduced and are not copyrighted. Single copies of this report will be available for a period of 90 days from the date of this report from the NIOSH Publications Office, 4676 Columbia Parkway, Cincinnati, Ohio 45226. To expedite your request, include a self-addressed mailing label along with your written request. After this time, copies may be purchased from the National Technical Information Service (NTIS), 5285 Port Royal, Springfield, Virginia 22161. Information regarding the NTIS stock number may be obtained from the NIOSH Publications Office at the Cincinnati address. Copies of this report have been sent to:

1. SEIU
2. UNICCO
3. Pro-Team
4. OSHA Region I

For the purpose of informing affected employees, copies of this report shall be posted by the employer in a prominent place accessible to the employees for a period of 30 calendar days.

Table 1 Summary Results of Biomechanical Analysis

Loading Condition	Muscle Force (lbs)	Disc Compression (lbs)
Upright/No BPVC	0	59
Upright/With BPVC	15	89
6.5° Flex/With BPVC	0	74
85° Flex/No BPVC	269	294
85° Flex/With BPVC	330	336

Table 2 Measurement of Average Direct Vibration (M/sec^2)

Freq (Hz)	x	y	z
0-5	0.100	0.491	0.098
120	0.001	6.704	0.006
167 or 193	0.003	-	-

Table 3 Measurement of Average Indirect Vibration (M/sec^2)

Condition	Frequency (Hz)/ Direction					
	.7-5			5-20		
	Direction			Direction		
	x	y	z	x	y	z
Standing	.006	.010	.006	.058	.096	.108
45° Flexion	.023	.006	.030	.102	.122	.120
90° Flexion	.008	.017	.007	.126	.143	.137
Working	.018	.007	.010	.132	.110	.269

Figure 1 Sketch of the BPVC showing the approximate dimensions.

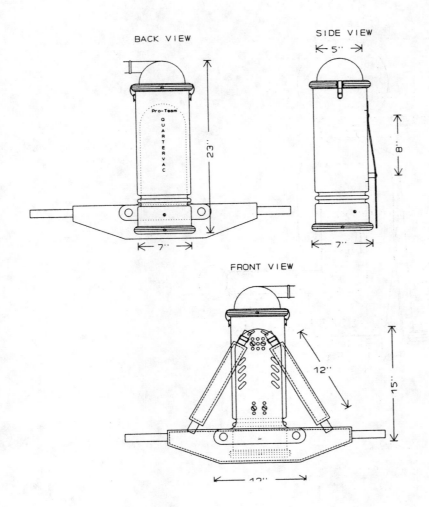

Figure 2 Sketch of horizontal cut through the L5/S1 intervertebral joint showing internal and external biomechanical forces.

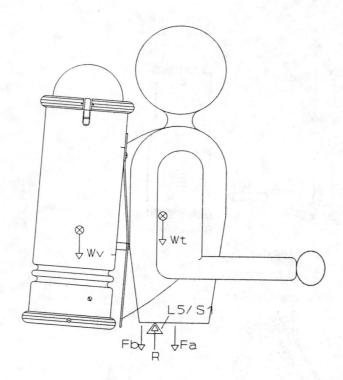

Figure 3 Free body diagram describing upright posture without BPVC.

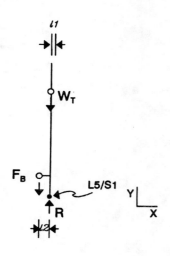

Figure 4 Free body diagram describing upright posture with BPVC

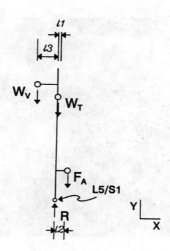

Figure 5 Free body diagram describing 6.5 degree forward flexion adjustment in posture to compensate for BPVC.

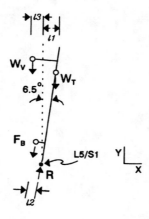

Figure 6 Video image of worker wearing the BPVC.

Figure 7 Free body diagram describing 85 degrees of forward flexion without BPVC.

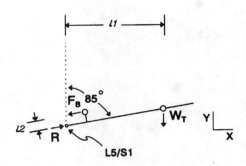

Figure 8 Free body diagram describing 85 degree forward flexion with BPVC.

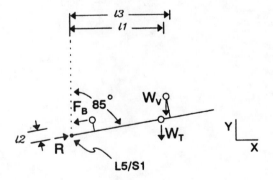

Figure 9 Schematic diagram showing placement of the accelerometers for the direct vibration test.

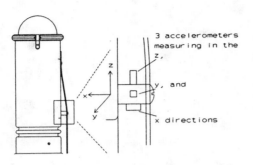

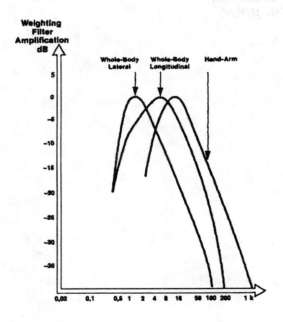

Figure 10 Frequency Weighting Curves provided by ISO

Figure 11 Graphical representation of results of skin temperature test.

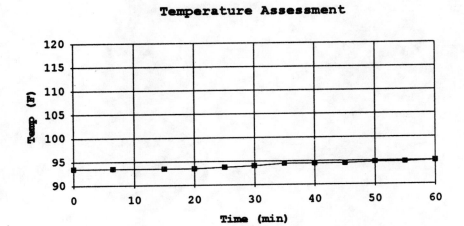

HETA 91-405-2340
August 1993
Big Bear Grocery Warehouse
Columbus, Ohio

NIOSH Investigators:
Ergonomic:
Vernon Putz-Anderson, PhD
Thomas Waters, PhD
Medical:
Sherry Baron, MD, MPH
Industrial Hygiene:
Kevin Hanley, MSPH, CIH

Physiologic Measurements

I. SUMMARY

On January 21, 1992, representatives from the National Institute for Occupational Safety and Health (NIOSH) undertook the first of four trips to the Big Bear Warehouse in Columbus, Ohio in response to a confidential request for a health hazard evaluation (HHE). The request was prompted by employee concerns about the potential hazards of repetitive lifting, excessive work pace, and heat stress associated with the job of "order selector." There were three objectives of this investigation:

1) Determine the prevalence and incidence of work-related musculoskeletal disorders (WMDs), particularly low back pain, associated with manual lifting activities of the order selectors in the grocery warehouse.

2) Document the presence of potential occupational hazards in the warehouse including repetitive lifting and heat stress.

3) Develop recommendations for preventing or reducing the physical stresses associated with repetitive lifting and heat stress.

During three separate visits over a six month interval, the NIOSH team collected information at the Big Bear work site to assess the extent and magnitude of the reported health hazards. During this and subsequent visits, we administered a questionnaires to the workers that included items designed to assess the workers' perceptions of physical workload and symptoms of musculoskeletal disorders. Occupational Safety and Health Administration (OSHA) logs and medical records were reviewed to determine the extent of the recorded injuries and lost time. Ergonomic analyses were conducted on the following dates:

1) On May 14, 1992, we conducted a preliminary analysis to determine variations in load weights and the types of stressful lifting postures that pose a risk to the worker.

2) On July 14, 1992, we returned and conducted a second-level of analysis on over 200 individual lifts. In addition to measuring load weights and postures, we used a lumbar motion monitor to measure trunk rotations. Measurements for assessing heat stress were also conducted on July 14.

3) On November 5, 1992, we examined the effects of lifting frequency and work duration as they affected oxygen use and heart rate.

The results of our medical evaluation confirmed that back injuries among selectors was the most important cause of lost workdays in the warehouse. Workers' compensation data provided by Big Bear for 1991 showed 22 cases of back sprain/strain among all the workers in the grocery warehouse (about 16 cases per 100 workers). On average, during the five years, back injuries among the selectors accounted for about 60% of all lost workdays in the grocerywarehouse.

Health Hazard Evaluation Report No. 91-405

A questionnaire was completed by more than 80% of current grocery selectors and showed that the work force was all male, who on average were younger, taller, and heavier than the majority of workers in the U.S. work force. More than 70% of the full time selectors reported significant physical discomfort in the region of the low back and 18% reported having a back injury during the previous year.

Based on a series of biomechanical and metabolic measurements of workers, NIOSH investigators were able to identify and quantify stressful work postures, motions, and levels of physical exertion that pose a significant risk of back injury. According to recognized criteria defining worker capability and accompanying risk of low back injury, the job of order selector at this work site will place even a highly selected work force at substantial risk of developing low back injuries based on the following findings from our work site analysis, which showed that workers are required to:

1) Lift loads that exceed recognized weight limits,
2) Lift loads that exceed recognized lifting rates,
3) Lift loads from stock locations that are either too low (near the floor), too high, or too great a horizontal distance from the spine resulting in excessive biomechanical strain,
4) Lift loads for work periods that sometimes exceed an 8-hr day, resulting in energy demands that exceed recognized capacities for a majority of workers.

Heat stress evaluation of this work location included measurement of the wet bulb globe temperature (WBGT) index, an assessment of the air velocity, and an estimation of the metabolic heat load of the task. Basic physiologic monitoring of the workers which included heart rate, and oxygen consumption was also used to comparatively establish the metabolic demand. Environmental monitoring revealed WBGT values that ranged from 76.5 - 79°F, with the peak dry bulb temperatures of 89°F. The metabolic heat load of the grocery selector job was determined to be moderate to high. Although the environmental heat conditions were not extreme, the metabolic demands of the job tasks caused the order selectors exposure to approach the heat stress (WBGT) criteria established by NIOSH and ACGIH. It is likely that on extremely hot and humid days, the heat stress criteria would be exceeded for the grocery selectors.

Health Hazard Evaluation Report No. 91-405

The objective of the present investigation was to determine if the job of order selector posed a substantive risk for development of musculoskeletal disorders, with particular reference to low back pain. The rate of OSHA 200 entries as well as workers' compensation claims for low back pain were elevated for order selectors despite the highly selected nature of this workforce.

Biomechanical and metabolic job analyses were performed to identify and quantify stressful work postures, heavy loads, and high frequency lifting motions. The order selectors at the grocery warehouse lift loads that are too heavy at excessive lifting rates from stock locations that are either too low, too high, or too far away from the body for work periods that can exceed an 8-hr day. According to recognized criteria defining worker capability and accompanying risk of low back injury, the job of order selector at this worksite poses a significant health hazard.

The ergonomic job analyses using a variety of criteria for identifying hazardous job tasks were in general agreement that order selecting is a hazardous activity with a substantial risk of low back injury. These hazards are not only the result of unique characteristics of the work methods or workplace layout, but also result from a combination of the amount of weight lifted per hour, number of lifts per hour, and lifting of objects from floor level and above waist height. Although the environmental heat conditions were not extreme, the metabolic demands of the job tasks caused the order selectors exposure to approach the heat stress criteria. Recommendations are provided in section VIII which include changes in work organization and methods.

Keywords: SIC 5411 (Grocery Stores, Warehouse), repetitive lifting, lumbo-sacral stress, back injury, biomechanical, physiologic, production standards, muscle fatigue, heat stress, wet bulb globe temperature index, metabolic heat load, convective heat exchange.

Health Hazard Evaluation Report No. 91-405

II. INTRODUCTION

On January 21, 1992, representatives from the National Institute for Occupational Safety and Health (NIOSH) visited the Big Bear Grocery Warehouse in Columbus, Ohio, in response to a confidential employee request for a health hazard evaluation. The request was prompted by employee concerns about the potential hazards of repetitive lifting, excessive work pace, and heat stress associated with the job of "order selection." During the initial meeting, NIOSH staff talked with the employer and worker representatives and discussed the objectives of the investigation and the level of effort required. The objectives of this investigation included:

1) Determining the prevalence and incidence of work-related musculoskeletal disorders (WMDs) associated with manual lifting activities of workers in the nonperishable grocery warehouse who perform the job of "order selector."

2) Documenting the presence of potential hazards in the warehouse, including heat stress and ergonomic issues related to repetitive lifting.

3) Developing recommendations for preventing or reducing the physical stresses associated with repetitive lifting and heat stress.

An interim report was sent to the company and to the requestors in March, 1992, which discussed preliminary findings and proposed future assessments.

III. BACKGROUND

A. Plant and Job Description

Big Bear is a retail grocery food supplier with warehouses located in Columbus, Ohio. Although the warehousing operation includes several buildings, this investigation was limited to warehouse 1, which contained all of the nonperishable grocery items. This area was chosen because the majority of selectors worked in that facility and it had a higher rate of reported injuries.

1. Workforce

The company employs, on average, about 145 both full- and part-time workers in Warehouse 1, 67 (46%) of whom select groceries each day. These workers are referred to as "order selectors." Both full-time and part-time workers average about 40 hours per week of work, although the contract allows for up to 10 hours of mandatory overtime. Part-time workers are employed on a temporary basis and are not covered by the union contract. Part-time workers make up 15-20% of the order selectors. The work schedule consists of two shifts that overlap during the mid afternoon. The company reported a very low turnover rate of full-time workers[1], but the average absence rate is 15-20%. Each new (part-time) employee is given one week of training and has a 30-day trial period. The company provides no "light duty" jobs for injured workers; they are allowed to return only when they are fully fit.

---

[1] The low turnover rate is in part a function of using a temporary work pool, which serves to screen employees.

Health Hazard Evaluation Report No. 91-405

2. Job Activity

The job of an order selector involves selecting cases of grocery items from supply pallets and loading the cases on an electrically-driven pallet jack. The selection or "picking order" is dictated by a computer-generated list that contains the items and quantities of groceries to be picked, the order for picking these items, and their locations (aisle and slot numbers). The cases of grocery items are located in supply pallets or slots on either side of the aisle. The grocery items are stored on wooden pallets located in either single or double deep slots, and may include a one or two item pick (two grocery items stored in the same slot at different levels). The total weight of the order and the total number of items per order may vary considerably.

According to data provided by the company from a recent 26 week period, on average, order selectors spend approximately 75% of their work time performing these tasks (range 44%-90%). The remainder of the time is spent on breaks, lunch, and doing a variety of other assigned tasks, including driving forklifts and doing maintenance work.

3. Job Cycle

A job sequence for the order selector, hereafter referred to as a job cycle, typically involves the following steps: (1) Walking or riding on an electric jack to the dispatcher window to receive a picking order for grocery selection; (2) driving to the empty pallet stacks and lifting or sliding two empty pallets onto the fork lift; (3) driving to the slot on the first picking label, walking to the grocery item to be picked, lifting the case from the slot, carrying it to the pallet jack, and lowering or lifting it on the pallet jack to build the load on the pallet jack; and finally, (4) peeling the label from the order and applying it to the case (item picked). This work sequence is continued until the entire order is picked. When the order is completed, the selector drives to the loading dock area and places the loaded pallets there. The order selector returns to the dispatcher office with the end sticker, receives his performance rating, and is given another order. This work sequence continues throughout the course of the workday.

4. Incentive System

A few years ago, the company had installed a "standard incentive program," developed by an industrial engineering consulting firm. Standards were set based on a time measurement system, amount of weight, movement time between items, and lift height. The goal of the standard incentive system was to establish a "fair amount of time" to do a work cycle. A work cycle was defined by the consulting firm as "the selection and filling of a 2-pallet cube." Each motorized jack could carry two pallets. Achieving 100% of the standard was defined as a "day's work," and was averaged over a week of work. An employee was disciplined for performance below 95% of the standard. To allow for conditioning, there was a four-week build-up period. The incentive consisted of increased money or time off and occurred when the worker exceeded the standard. Before the standards, they had used historical in-house data as standards, based on the number of cases handled per hour. The company representatives stated the old production-based standards were not cost-effective.

B. Incidence and Costs

The job of order selector has been previously identified as physically demanding, primarily because of the frequent and heavy manual lifting demands associated with these jobs (Garg 1986). One of the most frequent complaints associated with heavy manual lifting is back pain. As an occupation, "stock handlers" have an annual estimated prevalence of back pain caused by "activities at work" of 17.8%. More specifically, for males working in the wholesale grocery industry, the estimated annual prevalence of back pain caused by "activities at work" is 16.4%. This industry is one of the 15 industries with the highest prevalence of work-related back pain (Guo et al., 1993).

One study has shown that warehousemen averaged nearly 10 claims for workers' compensation per 100 workers during a given year (Klein et al., 1984). Moreover, others have shown that the majority of the back claims identify manual lifting as the primary cause (Bigos et al., 1986; Snook et al., 1978). Unlike many occupational diseases, these disorders do not wait until the worker is older to appear, but occur most frequently in otherwise young and healthy workers. The average age of workers filing compensation claims is 34 years. Costs per case in 1986 averaged over $6,800. In today's economy, the average cost can easily exceed $10,000 per case (Webster and Snook, 1990).

Workers' compensation data, supplied by the National American Wholesale Grocers' Association and the International Foodservice Distributors Association for the years 1990 to 1992, revealed that back sprains/strains accounted for 30% of all injuries for warehouse workers (NAWGA and IFDA, 1992). Data from the same report indicated that more than a third of all workers (34.6%) experience an annual injury in warehouse operations, accounting for an hourly cost of $0.61 per worker-hour. Moreover, manual lifting is identified as the cause of the back injury in 54% of the cases, followed by "push and pull" as a cause in only 15% of the cases. In summary, it is evident that grocery warehouse workers face a substantial risk of lifting-related low back disorders that include low back pain, overexertion, and strains and sprains.

IV. Evaluation Design and Methods

A. Medical Evaluation Methods

1. OSHA 200 Logs and Workers' Compensation

The OSHA 200 logs were obtained from the company for the period 1987-1991. These logs are the official report of occupational injuries and illnesses and are required by the Occupational Safety and Health Administration. Information from the logs was reviewed and rates of injuries and numbers of lost workdays were calculated. Data from workers' compensation claims for 1991 were also provided by the company.

2. Questionnaire

On July 14, a questionnaire was distributed to all Warehouse 1 grocery selectors who were at work that day on either the first or second shift. The questionnaire was completed by the workers during work hours. This questionnaire (Appendix A) included items that asked workers about the perceived physical workload of their job, symptoms of pain associated with musculoskeletal injuries, and whether they had experienced an injury during the previous year. Questions were also included concerning the overall workload and the workers' perceived control over their workload. A more complete description of these indicators is provided below:

Health Hazard Evaluation Report No. 91-405

Assessment of Perceived Physical Workload

The Borg scale was used to illicit an overall assessment of the perceived physical workload of the selectors' job. This scale consists of a 15-point numerical list, anchored by adjectives describing increasing levels of physical effort (Question 12, Appendix A). The Borg scale was initially developed through laboratory experiments using exercise bicycles and has subsequently been used at the worksite to assess the perceived physical effort of persons performing manual tasks. Studies have shown a good correlation between perceived workload and objective measures of physiologic workload such as heart rate (Borg 1982, Borg 1990).

Assessment of Reported Discomfort

Several investigations have used questionnaires to determine the prevalence of musculoskeletal disorders among working populations. A particularly descriptive method for determining the location and severity of complaints is the Corlett-Bishop (1976) body parts map diagram (Question 13, Appendix A). A number of studies have documented the relationships between complaints of discomfort and inadequate ergonomic work conditions. These questionnaires are useful in identifying which parts of the body are under the greatest stress. (Kuorinka et al. 1987, Silverstein et al. 1986, Viikari-Juntura 1983)

Assessment of Injuries and Missed Workdays

Workers were asked about injuries at work and lost workdays due to those injuries. Although these cases probably represent the more severe problems, they provide another indicator of the magnitude of the problem.

Employees Perception of Job Demands and Job Control

A series of questions was included to determine selectors' perception of their job demands and control (Pages 3-5, Appendix A). These questions were chosen based upon a decision latitude model of job stress. This model suggests that a combination of high-job demands and low-job control will produce high job strain and could lead to problems such as stress and job dissatisfaction (Landsbergis, 1988). Questionnaires containing similar questions have been administered to thousands of workers employed in a variety of occupations, thus allowing a comparison of this job to a range of other occupations (Hurrell and Linstrom, 1992).

Control: Control was measured using a scale factor analytically derived (McLaney and Hurrell, 1988) from work originally conducted by Greenberger (1981). This seven item scale (Question 18, Appendix A) measures task related control and includes items assessing individual control over the variety, order, amount, pace, and quality of work performed. The scale has been shown to be highly internally consistent (McLaney and Hurrell, 1988)

Quantitative Workload: This four item scale (originally developed by Caplan et al., 1975) contains items which assess the quantity of work required of the job incumbent (Question 15, first 4 items, Appendix A). The scale has also been shown to have high internal consistency (Caplan et al., 1975).

Health Hazard Evaluation Report No. 91-405

B. Ergonomic Assessment

   1. Risk Factor Identification

   To assess the musculoskeletal stresses associated with repetitive manual lifting among Warehouse 1 order selectors, the NIOSH team collected information on the following factors, each of which is a recognized risk factor for overexertion and low back pain:

   1) Load or weight of the objects to be lifted (Chaffin and Ayoub, 1975);

   2) Posture of the worker in reference to the position of the load to be lifted (Chaffin et al., 1977);

   3) Dynamics of the lifting motion that affect spinal forces, i.e., dynamic trunk rotations (Marras et al., 1993); and

   4) Frequency and duration of manual lifting activities, i.e., the temporal pattern of manual lifting, including work-rest cycles (Chaffin et al., 1977).

   A main advantage of organizing our data collection by risk factor is to ensure that the (1) appropriate information is collected, and (2) the information that is collected can be readily evaluated against known criteria to determine what constitutes low-risk or high-risk jobs for a majority of healthy workers.

   2. Established Criteria

   Two sets of criteria are commonly used to evaluate the potential risk associated with the manual lifting tasks: biomechanical and metabolic (NIOSH 1981). Biomechanical risks refer to the mechanical stress on the musculoskeletal system resulting from lifting, while metabolic risks refer to the physiological stress imposed by the workload of the lifting job. Researchers have developed hazard assessment models which utilize these risk factors to estimate the likelihood that a particular lifting job poses a significant risk of a work-related injury and for low back pain (Waters et al., 1993, Marras 1993).

   Biomechanical Evaluation of Load and Posture.

   For this initial analysis, we conducted two separate evaluations. Information on load weights and body postures was systematically recorded for **five representative lifting tasks that were judged by both the workers and investigators as having a high potential for injury.** Each of the five lifting tasks are shown in Figure 1. Each example illustrates a different sample of stressful lifting postures observed by the investigators. (The pictures were taken directly from the video-tapes of the sampled lifting tasks.)

   1) Lifting Task 1: 30 lb load, trunk flexion, no twist.
   2) Lifting Task 2: 38 lb load, long reach, small twist.
   3) Lifting Task 3: 42 lb load, trunk flexion, high twist.
   4) Lifting Task 4: 38 lb load, long reach, high twist.
   5) Lifting Task 5: 58 lb load, shoulder high reach, twist.

   By examining a sample of the potentially most hazardous jobs first, the overall severity of the hazards can be estimated. This information is useful in planning the sampling scheme and measurement precision required for subsequent in-depth analyses.

Health Hazard Evaluation Report No. 91-405

Procedure: Two experienced male grocery selectors were randomly chosen to participate in this phase of the investigation. Both of the workers, referred subsequently to as participants, were healthy and conditioned for work. Both participants were informed of the investigative procedures before data collection began. Participant 1 performed Tasks 1-4, and Participant 2 performed Task 5 (See Figure 1). Each participant was instructed to perform the lifts using the same technique he would use when actually selecting a grocery order.

For those lifts which were selected for analysis, each participant was asked to momentarily hold his position at the lift-off point and set-down point. During the approximate 10-15-second interval, the investigator was able to record 15 angular values with the aid of an electro-goniometer (Lafayette Instruments, Inc., Model # 35). These data values serve as the input for the Michigan 3D Static Strength Prediction Program (SSPP). The Michigan 3D SSPP model provides estimates of lumbo-sacral (L5/S1) disc compressive forces as well as information on the muscle strength requirements which a person needs to perform the designated lift (Chaffin and Andersson, 1991). The model was used to estimate the percentage of the working male population which would have sufficient strength at the hip joint and at the torso to lift each of the objects. These two body parts were chosen because they are most closely associated with the development of low back pain.

Videotape and still photographs also were made for each lift to assist the analyst in interpreting the measured joint angles for the analysis. In addition, the measurements were recorded using the electro-goniometer and a standard retractable tape measure. These measurements were needed to apply the NIOSH lifting equation (Appendix B).

<u>Trunk Motion Measures</u>

The purpose of the trunk motion evaluation was to further define the level and extent of the biomechanical hazards associated with the job of order selector. Because the lifting jobs at the worksite were repetitive and paced through an incentive system, a second-level analysis was designed to focus on the dynamic properties of manual lifting to determine the probability that the lifting tasks posed a significant risk for low back injury. To accomplish this, we used a commercial Chattanooga Corporation Lumbar Motion Monitor™ (LMM) that was worn by the worker (Marras et al. 1993). The LMM is capable of measuring the asymmetrical pattern of tri-axial spinal motion (rotation) around the L5/S1 intervertebral joint that occurs as a lift is performed.

Procedures: Three experienced male grocery selectors from the same warehouse were randomly chosen to participate in this phase of the investigation. Each participant was informed of the procedures before we began the data collection.

Each of the three participants was fitted with the LMM and monitored for one complete job cycle. (A job cycle was defined as a complete order selection.) For each lift and lowering motion, the participant was asked to identify the specific item to be selected and where it was to be placed on the pallet. Thus, each item was categorized into one of nine possible types of lift and lowering motions at three vertical heights for both the origin and destination of the lift. Vertical height was defined as either low, medium, or high (L,M, or H), with low defined as below 30 inches, medium defined as between 30 and 50 inches, and high defined as above 50 inches. Each participant was also video-taped as he performed order selections. The video tape was used as an independent audit trail to verify which grocery items were selected and how they were lifted. To ensure an adequate sample of lifts and lower motions for each of the three vertical height categories, approximately 200 lifting tasks were evaluated.

Metabolic Measures

To characterize the metabolic risk factors during the third and final assessment visit, the NIOSH team used three procedures to assess the energy demands of order selection: (1) oxygen measurement, (2) heart rate monitoring, and (3) energy expenditure modelling. Oxygen measurement provides a relatively objective assessment of the energy demands posed by the work load of the job, whereas heart rate monitoring and metabolic modeling are indirect procedures for assessing metabolic load, inherently less accurate, but somewhat easier to use.

Procedure: Three male grocery selectors were randomly chosen for this phase of the investigation. (For each phase of the investigation, different individuals participated.) All participants were informed of the procedures before the data were collected.

For our testing trials, a management representative selected the grocery orders for each selector-participant. The order was judged to be of average size and difficulty. Each participant was fitted with an Oxylog portable oxygen consumption meter (Morgan Instruments, Inc.) and a Polar portable heart rate monitor (Polar USA Inc.) (Ballal and Macdonald 1982). Heart rate data were collected from a combination electrode-transmitter band that was worn on the chest and a watch-like receiver that was worn on the wrist. The receiver was able to store more than 8-hours of heart rate data for subsequent down-loading into a computer for analyses. The oxygen consumption values were displayed on the unit's display and manually recorded by the investigator every 5 minutes during the selection cycle. Oxygen consumption and heart rate were allowed to stabilize for approximately five-to-ten minutes before the grocery order was selected. Each of the three participants were instructed to work as they normally would and to maintain a work pace equivalent to 100% of the existing performance standard.

To assist in the metabolic analysis, a printout of the order (i.e., items to be selected) was provided by management. This showed the total number of items selected, the weight, volume (size) of each item and location of each item. We also videotaped the order selections to allow us to conduct an independent task-based metabolic assessment using a model developed by Garg and others (Garg et ai., 1978).

C. Heat Stress Evaluation

Environmental heat assessment of this work location included measurement of the wet bulb globe temperature (WBGT), an assessment of the air velocity, and an estimation of the metabolic heat load of the work task(s). Basic physiologic monitoring of the workers, which included the previously discussed heart rate and oxygen consumption measurements, was also utilized to determine the metabolic heat demand.

On May 14, 1992, a "baseline" heat stress evaluation was conducted to evaluate the conditions associated with a typical moderate weather day (of spring and autumn). On July 14, 1992, the heat stress measurements were repeated on a warm summer day to determine the exposure conditions more typical of this season. During each of these surveys, environmental heat measurements were obtained at twelve locations spatially spread throughout the warehouse. These measurements were obtained near the beginning of the first shift (early afternoon), as well as mid-way through the work shift (late afternoon/early evening). A stationary site (aisle 56, slot #3810), located in the center of the warehouse, was also selected for collecting measurements at five-minute intervals during the majority of the first shift (approximately 12 noon to 6 PM). In addition, a heat stress meter was mounted on a pallet jack, and data were logged at short intervals during order selecting to evaluate the cooling effect, if any, due to the wind induce by riding the pallet jack.

Health Hazard Evaluation Report No. 91-405

Environmental measurements were obtained using a Reuter Stokes RSS 211D Wibget® heat stress meter manufactured by Reuter Stokes, Canada. This direct reading instrument is capable of monitoring dry bulb, natural (unaspirated) wet bulb, and black globe temperatures in the range between 32° and 200°F, with an accuracy of ± 0.5-1.0°F. This meter also computes the indoor and outdoor WBGT indices in the range between 32° and 200°F. Measurements were collected about four feet from the floor after the meter was allowed to stabilize. A Veloci-Calc™ thermo-anemeometer, manufactured by TSI, was used to collect air velocity measurements with a read-out accuracy of ± 2.5%.

The heart rate and oxygen consumption measurements obtained during the ergonomic and physiologic assessment allow for an accurate determination of the metabolic rate required to perform the measured job tasks. However, the difficulty level of selecting a given grocery order can vary significantly, and only a limited number of heart rate and oxygen consumption measurements were collected. Therefore an estimation of the metabolic rates of this work was also performed using energy expenditure tables and the guidelines provided in <u>Occupational Exposure to Hot Environments, Revised Criteria 1986,</u> and <u>Threshold Limit Values for Chemical Substances and Physical Agents</u> (NIOSH, 1986; ACGIH, 1992). Using this method, the average energy expenditure for a "standard" male worker (body weight 70 kilograms; body surface 1.8 square meters) can be calculated utilizing basal metabolism, and specific task analysis information regarding body position, movement and type of work. Table 12 lists the average energy requirements for the task analysis components. Assessment of the metabolic heat demand of the job task(s) is essential to allow one to apply the appropriate WBGT evaluation criteria to the observed environmental conditions. It is important to note that errors in estimating metabolic heat from energy expenditure tables are reported to be as high as 30% (NIOSH, 1986). Because of the error associated with estimating metabolic heat, NIOSH recommends using the upper value of the energy expenditure range reported in Table 12 to allow a margin of safety. The metabolic rates obtained from each of these methods were compared and the most protective estimate was used to establish the appropriate heat stress exposure criteria.

Heat stress is defined as the total net heat load on the body with contributions from environmental sources and from metabolic heat production (Dukes-Dobos, 1981). Four factors influence the exchange of heat between the human body and the environment. These are: (1) air temperature, (2) air velocity, (3) moisture content of the air, and (4) radiant temperature. The fundamental thermodynamic processes involved in heat exchange between the body and its environment may be described by the basic equation of heat balance:

$$S = M - E \pm R \pm C$$

where:

S = the change in body heat content (heat gain or loss);
M = metabolic heat gain associated with activity and physical work;
E = heat lost through evaporation of perspiration;
R = heat loss or gain by radiation (infrared radiation emanating from warmer surfaces to cooler surfaces);
C = heat loss or gain through convection, the passage of a fluid (air) over a surface with the resulting gain or loss of heat.

Under conditions of thermal equilibrium (essentially no heat stress), heat generated within the body by metabolism is completely dissipated to the environment, and deep body or core temperature remains constant at about 98.6°F (37°C). When heat loss fails to keep pace with heat gain, the body's core temperature begins to rise. Certain physiologic mechanisms begin to function in an attempt to increase heat loss from the body. First, the body attempts to radiate more heat away by dilating the blood vessels of the skin and subcutaneous tissues and diverting a large portion of the blood supply to the body's surface and extremities. An increase in circulating blood volume also occurs through the withdrawal of fluids from body tissues. The circulatory adjustments enhance heat transport from the body core to the surface. If the circulatory adjustments are insufficient to adequately dissipate excessive heat, sweat glands become active, spreading perspiration over the skin to remove heat from the skin surface through evaporation.

There are a number of heat stress guidelines that are available which are intended to protect against heat related illnesses such as heat stroke, heat exhaustion, heat syncope, and heat cramps. These include but are not limited to the wet bulb globe temperature (WBGT), Belding-Hatch heat stress index (HSI), and effective temperature (ET) (Yaglou et al, 1957; Belding et al, 1955; Houghton et al, 1923). The underlying objective of these guidelines is to prevent workers' core body temperature from rising excessively. The World Health Organization has concluded that "it is inadvisable for deep body temperature to exceed 38°C (100.4°F) in prolonged daily exposure to heavy work" (WHO, 1969). Many of the available heat stress guidelines, including those proposed by NIOSH and the American Conference of Governmental Industrial Hygienists (ACGIH), also use a maximum core body temperature of 38°C as the basis for the environmental criteria (NIOSH, 1986; ACGIH, 1992).

Both NIOSH and ACGIH recommend the use of the WBGT index to measure environmental factors because of its simplicity and suitability in regards to heat stress. The International Organization for Standardization (ISO), the American Industrial Hygiene Association (AIHA), and the Armed Services have published heat stress guidelines which also utilize the WBGT index (ISO, 1982; AIHA, 1975; U.S. Dept of Defense, 1980). Overall, there is general conformity of the various guidelines; hence, the WBGT index has become the conventional technique for assessment of environmental conditions in regards to occupational heat stress.

The WBGT index takes into account environmental conditions such as air velocity, vapor pressure due to atmospheric water vapor (humidity), radiant heat, and air temperature, and is expressed in terms of degrees Fahrenheit (or degrees Celsius). Measurement of WBGT is accomplished with an ordinary dry bulb temperature (DB), a natural (unaspirated) wet bulb temperature (WB), and a black globe temperature (GT) as follows:

$$WBGT_{in} = 0.7 (WB) + 0.3 (GT)$$

for inside or outside without solar load,

Or

$$WBGT_{out} = 0.7 (WB) + 0.2 (GT) + 0.1 (DB)$$

for outside with solar load.

Originally, NIOSH defined excessively hot environmental conditions as any combination of air temperature, humidity, radiation, and air velocity that produces an average WBGT of 79°F (26°C) for unprotected workers (NIOSH, 1972). However, in the revised criterion for occupational exposure to hot environments, NIOSH provides Recommended Exposure Limits (RELs) of WBGT environmental conditions in accordance with specific work-rest cycles and metabolic heat production (NIOSH, 1986). NIOSH has developed two sets of recommended limits; one for acclimatized workers [recommended exposure limit (REL)], and one for unacclimatized workers [recommended action limit (RAL)]. Refer to Figure 3 for the diagrams describing the RELs.

Similarly, ACGIH recommends a Threshold Limit Values® (TLV) for environmental heat exposure permissible for different work rest regimens and work loads (metabolic heat) (ACGIH, 1992). The NIOSH REL and ACGIH TLV criteria assume that the workers are heat acclimatized, are fully clothed in summer weight clothing, are physically fit, have good nutrition, and have adequate salt and water intake. Additionally, they should not have a preexisting medical condition that may impair the body's thermoregulatory mechanisms. Modifications of the NIOSH and ACGIH evaluation criterion should be made if the worker or conditions do not meet the previously defined assumptions.

It is important to distinguish the difference between the qualifying terms "high, moderate, & low," in regards to metabolic rate when applying it to a heat stress evaluation or a physiologic (ergonomic) assessment. The use of metabolic rate in a heat stress evaluation is necessary to provide an approximation of the bodily heat production as a result of the manual work performed, whereas the metabolic rate of a physiologic assessment is used as an indicator of muscle fatigue potential.

V. RESULTS

A. Medical

1. OSHA 200 Logs and Workers Compensation Data

Table 1 summarizes the rates of all injuries and back injuries alone among the selectors in Warehouse 1 between 1987 and 1991. These rates were based upon a population of 67 selectors (54 full-time and 13 part-time) as indicated by the company. The data provided in Table 1 also shows the total number of lost workdays each year due to back injuries and the average number of lost workdays for a back injury case. On average, during the five years, back injuries among the selectors accounted for about 60% of all lost workdays in Warehouse 1, including all jobs and all types of injuries.

According to the Big Bear risk management loss-prevention department, the rate of total OSHA 200 incidents for Warehouse 1 for 1990 was 32.2 lost-time and 12.9 no-lost-time per 200,000 manhours (100 40-hour per week workers). The Bureau of Labor Statistics (BLS) reports that for the Standard Industrial Classification (SIC) code for warehousing (422), which includes grocery warehouses, the average rate for 1990 was 8.2 lost-time and 7.2 no-lost-time incidents per 200,000 manhours. This shows that the rates are higher than the warehousing industry in general and that most of the excess is in lost-time injuries.

Workers compensation data provided by Big Bear for 1991 showed 22 cases of back sprain/strain among all workers in Warehouse 1 (about 16 cases per 100 workers). One investigation found that warehousemen averaged nearly 10 claims per 100 workers in 1984 (Klein et. al., 1984), whereas the annual incidence rate for lifting injuries for the total industrial population is about 2%.

2. Questionnaire

Eighty-one percent of all selectors were present the day the questionnaire was administered, 20 first-shift selectors and 34 second-shift selectors. Of those present, 48 (89%) completed the questionnaire. Of the 48 selectors, 10 were "part-time" temporary workers who had been hired within 6 months of the questionnaire. All questionnaire results are presented with the full-time and part-time workers separated.

Health Hazard Evaluation Report No. 91-405

Demographics

Table 2 shows the demographics of these groups. The workforce was 100% male, relatively young by national standards, with an average age of 29 years as compared with the national average age of the workforce of 35.8 years (BLS, 1992). Moreover, the average work experience in warehouse work was less than 6 years. In terms of physical size, the workers were above average in both height (181.1 cm) and weight (86.7 kg), in the 90th, and 60th percentile, respectively for American male workers.

Reported Injuries and Missed Workdays

Among the 10 part-time selectors, 3 (30%) reported missing work during the past year due to an injury at work. One injury involved the leg, one the elbow, and a third was unspecified. The average number of missed workdays was four. Among the 38 full-time selectors, 19 (50%) reported one or more injuries in the past year. Table 3 shows a detailed description of those full-time workers' reported injuries. Back injuries were the most common injury, with 18% of full-time selectors reporting a back injury during the previous year.

Although still substantial, the rate of back injuries from Table 3 was lower than that found on the OSHA 200 logs, which includes both full-time and part-time workers. This discrepancy can be explained in several ways: 1) workers may forget about minor injuries, 2) the full-time workforce is the most experienced workforce and may have adopted work techniques which protect against injuries, or 3) the full-time workforce represents a select group of workers who may be less likely to develop an injury than all workers. Nonetheless, these findings provide additional evidence that back injuries are a substantial cause of morbidity among grocery selectors.

Reported Symptoms

The body discomfort map shown in Appendix A, Question 14 was simplified into four body part areas for purposes of analysis: neck/shoulder, back, hand/arm, and lower extremity. Figure 2 shows which area were included in each category. A worker was considered to have significant discomfort if he reported pain that was "very or extremely uncomfortable," the top half of a 4-point scale. Table 4 shows the rate of "very or extremely uncomfortable pain" in each of the four-body part areas. For the full-time workers, over 70% reported significant back pain in the past year, while the rates for the other body parts ranged between 24 and 37%. These finding suggest that in addition to the 20-30% of workers with recorded back injuries, there are a substantial proportion of selectors who continue to work with substantial back pain.

Borg Scale

The full-time workers reported an average Borg score (Appendix A, Question 12) of 17 (very hard physical effort), with a range of 13-20. The part-time workers reported an average of 15 (hard physical effort), with a range of 11-19. These findings correlate well with those found using heart rate and oxygen consumption monitoring, described below.

Job Demand and Control

All of the items included in each of the two scales were combined and overall average demand and control scores were computed. Only the 38 full-time workers were included in this analysis. These results were then compared to the results from a previous NIOSH investigation of 2300 Maine public employees working in a wide range of occupations. The average demand score for the warehouse workers was 3.9 (fairly often demanding), while

Health Hazard Evaluation Report No. 91-405

the average for the Maine public employees was 3.4 (sometimes-fairly often demanding). The biggest difference between these two groups was in their reports of how frequently their job requires them to work very fast or very hard. The average score for these two indicators alone for the warehouse and Maine public employees were 4.2 and 3.2, respectively. The average control score for the warehouse workers was 2.0 (little control), while the average for the public employees was 3.7 (moderate-much control). All of these differences were statistically significant at at least the $p = .001$ level.

These finding suggest that order selecting is a high-demand low-control job, as compared to a large diverse group of public employees. In particular, the selectors reported a much lower level of control over their job activities. During informal interview with selectors, a common concern was the effect of the work standards on the workers ability to control the pace and content of their job. Researchers have shown that this combination of high demand and low control could lead to problems such as stress and job dissatisfaction (Landsbergis, 1988).

B. Ergonomic

1. Biomechanical Data

a. NIOSH Lifting Equation

Table 5 displays the results of the evaluation of five sample lifting tasks that are shown in Figure 1. Information on the use of the NIOSH revised equation and the complete analyses for the five tasks is found in Appendix B. The column labeled RWL in Table 5 refers to the Recommended Weight Limit, and the column labeled LI refers to the Lifting Index. The LI is computed by dividing the actual load by the RWL. Lifting tasks recognized by NIOSH as posing a low level of risk for the majority of workers will have LI $<$ 1.0. A LI that exceeds 1.0 increases the risk of low back injury for an increasing number of individuals. Many researchers believe that a LI greater than 3 poses a risk of back injury for the majority of workers (Waters et al, 1993).

Three of the five lifting tasks (Tasks 2-4) shown in Table 5 required horizontal reach distances that exceed the maximum allowable distance of 25 inches. In order to make the calculation comparable for all 5 tasks, the maximum distance (25 inches) was used in the calculation. For all five of the tasks, the weight lifted (L) far exceeded the RWL, which resulted in LI values ranging between 4.2 and 8.0, respectively.

If the basic design of the warehouse racks and the size and weight of grocery cases remain the same, only two of the factors in the lifting equation can readily be reduced: 1) the horizontal distance between the load and the body and 2) the frequency of lifts. Therefore, the LI can be brought closer to 1 by either moving the loads closer to the body during a lift and/or decreasing the required lifting rate.

b. Michigan 3-D Model

Table 6 displays the results of the evaluation of the five tasks shown in Figure 1 using the University of Michigan's 3D SSPP model. Four of the five tasks evaluated had a disc compression force at the lumbo-sacral disc greater than 770 pounds (3.4 kN). Compressive force values of 3.4 kN and larger have been identified with increasing rates of reported low back pain and lost time (Herrin et al. 1986). This value also serves as the maximum compressive force level reported as the "biomechanical criterion" for the revised NIOSH lifting equation (Waters et al., 1993).

Those four lifting tasks that produced maximum compressive force levels greater than 770 lbs, based on the available evidence, would pose a significant risk to the majority of workers for the development of low back pain or overexertion injury. The only lifting task with a disc compression value below 770 pounds was Task 4.

Only a fraction of the U.S. male workforce would have the strength capacity at both the hip and torso necessary to perform the five lifting tasks with the attendant postures. In fact, according to the model, only 16 percent of the male working population would have the hip strength capability to perform Task 3, and only about 50% of the workforce would have the hip strength capability to perform Tasks 4 or 5 (52% and 55%, respectively).

Table 6 also provides data on shear force and torsion. Shear force is defined as the transverse force applied perpendicular to the vertical axis of the spinal segment. Resultant shear force is defined as the sum of the sagittal and frontal plane shear force components. Although there are no published criteria on spinal shear forces, the results displayed in Table 6 indicate that substantial shear forces exist for all five lifts evaluated.

Spinal torsion is defined as the magnitude of the twisting moment about the vertical axis of the spinal segment. When someone twists to pick up a load, the musculoskeletal system must develop greater force on one side of the spine than on the other. These asymmetric muscular contractions between the right and left side of the body usually result in torsional moments.

Research evidence suggests that increases in both shear force and spinal torsion, which accompany asymmetric loading, significantly increase the risk of low back disorders, especially when lifting heavy loads (Andersson, 1981; Majora, 1973). For the present, there are limited data to predict the exact relationship between combinations of shear force and torsional levels and the risk of musculoskeletal injury. However, the results displayed in Table 6 seem to indicate that asymmetric loading is present in these five tasks, especially for Task 2. In general, asymmetric loading is associated with lower levels of acceptable lifting capacities and an increased risk of low back injury (Garg and Badger, 1986; Bean et al., 1988).

c. Dynamic Trunk Motion Analysis

Lift Distribution

Table 7 provides a tabulation of the origin and destination of the lifts that were evaluated using the LMM. Of the 306 lifts that were performed by the participants during the measurement period, the LMM successfully recorded data on 216 (70%). From Table 7, it is evident that the 216 lifts were fairly evenly distributed with regard to origin and destination height, with 37, 28, and 35 percent originating in the low (below 30 inches), medium (30-50 inches), and high (above 50 inches) range, respectively, and 23, 37, and 40 percent ending in the low, medium, and high range, respectively. Lifts that both originate and end in the medium height range are considered desirable from a biomechanical perspective. However, only 7.4% of 216 lifts observed during our visit were in this category. This finding indicates that the majority of the tasks required lifting motions that began or ended either high or low.

To simplify the data, all lifts were classified in one of three categories: 1) low, if either the origin or destination was below 30 inches (53% of all lifts); 2) medium, if both the origin and destination were between 30 and 50 inches (7% of all lifts); and 3) high, for all other lifts (40% of all lifts). The lumbar motion monitor was then used to calculate the average trunk motion for all of the lifts in each of the three categories.

Health Hazard Evaluation Report No. 91-405

Marras et al. (1993) have developed a model using the LMM which, based on an analysis of over 400 industrial jobs, found that five variables were most effective in predicting which job titles were associated with a high rate of back injuries reported on the OSHA 200 logs. These five variables include three trunk motions which were obtained from the LMM: (1) the peak lateral velocity of the trunk, (2) the average twisting velocity of the trunk, and (3) the peak sagittal flexion angle of the trunk (forward bending angle). We computed the remaining two workplace variables needed for the model: (1) the average lifting rate, and (2) maximum L5/S1 moment, which is the product of the average horizontal distance between the worker and the load and the weight of the load.

Table 8 shows the results of these five variables for each of the three lifting categories. Using the Marras model, all three types of lifts would have a high probability of being included in a high risk category for low back injury.

Lifting Rate/Maximum Moment

The average lifting rate for the 216 tasks was 4.1 lifts/minute (246 lifts/hour). Load weights were obtained for 155 of the 216 lifts that were analyzed with the LMM. The average weight of the load for those lifts was 30.4 pounds [135 Newtons (N)]. The horizontal distance of the load from the spine was obtained for 141 of the 216 lifts analyzed with the LMM. The average horizontal distance at the origin of the lift for those lifts was 27.0 inches [0.69 meters (M)]. Thus, for an average lift, the estimated peak static spinal moment was 821 in•lbs (93 N•M).

The values shown in Column 3, Table 8 are based on the same data set, but were combined and averaged by lift height. For the low height range, below 30 inches, the maximum moment averaged 92 N•M; for the medium height, 30 - 50 inches, the maximum moment averaged 126 N•M; and, for heights above 50 inches, the maximum moment averaged 87 N•M. Maximum moments of this magnitude are likely to generate lumbo-sacral disc compressive forces well above the value of 3400 N, chosen as the upper limit of the biomechanical criterion for the revised NIOSH lifting equation (Waters et al, 1993).

Rotation Angle.

As the height of the lifts increased, there was a significant decrease in the average peak sagittal lumbar flexion angle ($p = 0.047$). The average peak (maximum) sagittal flexion recorded for the low, medium, and high categories was 56, 38, and 28 degrees, respectively (Column 4, Table 8). There were no significant differences between frontal and transverse ranges of motion.

The average peak transverse velocity for the low, medium, and high height categories was 42.6, 46.2, and 43.2 degrees/second, respectively. In a recent paper by Marras et al. (1993), the <u>average</u> twisting (transverse) velocity during a dynamic activity was found to be a better predictor of injury risk than the <u>peak</u> twisting velocity. Therefore, the average transverse rotation velocity also was determined for each lift. The means for the average transverse rotation velocity for each of the three height categories were 5.9, 5.8, and 6.5 degrees/second for the low, medium, and high height categories, respectively (Column 5, Table 8).

As with the range of motion variable, there was no significant difference in the average peak frontal or transverse velocity between the three height categories. The average peak lateral velocity was 31.0, 33.3, and 26.0 degrees/second, for the low, medium, and high category, respectively (Column 6, Table 8).

Health Hazard Evaluation Report No. 91-405

2. Metabolic Data

Table 9 displays a summary table of results for evaluating the metabolic demands for three cycles of order selection. The table provides information on the total cases per order, total weight per order, allowed time per order, average weight handled per minute, worker's performance index, measured metabolic working rate, measured average heart rate, and predicted metabolic rate. The length of time required to complete the order varied from 18 to 31.5 minutes, depending on the size of the order and the worker's pace.

During the monitored work cycles, all three participants exceeded the criterion for energy expenditure (5.0 kcal/min, as measured by oxygen consumption, and predicted using Garg's metabolic model). The aerobic work load, which ranged from 5.4 kcal/min to 8.0 kcal/min in our sample, is typically defined as "heavy" to "very heavy" work for individuals of average capacity (Eastman Kodak, 1986, Table 26-24). Average heart rates over the duration of the selection cycle for two of the three workers also exceeded the criterion of 110 beats per minute, recognized as acceptable for workers (Astrand and Rodahl, 1986). Average heart rates for younger, highly conditioned workers can be lower than what is found among the general working population.

Based on the metabolic assessments of three workers and the assessment of energy demands required to perform the job of order selector, these jobs pose a significant risk for overexertion injury from excessive physical fatigue for the majority of healthy workers who would perform this type of job. This conclusion is based on the energy demands of the jobs, which exceeded 5 Kcal/min.

3. Workplace Layout Analyses:

Two different types of racks or slots for storing the grocery items are used in Warehouse 1: (1) a one-level pick or single tier (walk-in slot, double deep) that was approximately 75" high by 101" wide and 105" deep, and comprised about 80% of the slots, and (2) a double tier (2-level pick) that also was 101" wide by 51" deep, but had an opening height of 98." The lower tier opening provided a 46" high clearance, and the upper tier clearance height was also 46." The two tiers were separated by a 5" shelf. The pallets on which the groceries were stored added another 3 - 4," which resulted in reduced head clearance.

Most workers under 6' were able to stand up in the walk in-slots, whereas with the double-tier slots workers typically had to squat or bend at the waist to reach the items in the lower slot. Workers were observed kneeling and crawling to reach some objects in the lower racks. Selecting grocery items from the top tier, located at a height of 51" could also pose a stress for shorter workers, since 51" is only 7" below the shoulder height of an average male. This effort is increased when the workers have to reach for objects which are located further back in the slot. Again, workers were observed climbing on cases in the lower tier to reach items in the upper tiers, particularly if they were not near the front edge.

Pallet heights on the jack were observed to be as high as 90" near the end of the selection order. Workers were observed climbing on their jacks to build their load to accommodate the size of the order. This poses a risk of falling and produces high levels of shoulder stress to maneuver heavy items above one's head. Workers were also observed having to hold objects in place on the pallet as they organized the fitting of boxes on the pallet.

Health Hazard Evaluation Report No. 91-405

C. Heat Stress

The heat stress potential that the grocery selectors were exposed to was evaluated in both the Spring and Summer seasons. During this evaluation, WBGT measurements were collected at a stationary location (in the center of the warehouse) as well as throughout the building. Environmental heat (WBGT) measurements were also obtained on a pallet jack when it was used for selecting orders.

The WBGT data collected during the Spring and Summer surveys are presented in Table 10 and Table 11, respectively. A total of 12 locations were selected, spatially spread throughout the warehouse to include sample locations in the front near the loading dock, in the back close to the axial fans as well as in central zones of the building. A meter, which logged heat stress data at 5-minute intervals (through most of the first shift), was also positioned in Aisle 56, slot 3810, in the center of the warehouse.

The purpose of the Spring survey was to establish the baseline conditions that the warehouse workers experienced during mild weather of spring and fall. During this evaluation, the ambient temperature (dry bulb) ranged from 70°F around noon to 62°F shortly after 6:00 PM. The interior heat conditions on May 14, 1992, were also quite mild and reasonably consistent throughout the building. The DB measurements ranged from 73 - 77°F, with WBGT measurements that ranged from 62 - 65°F. Relative humidities inside and outside the warehouse were consistently in the mid 30%. The air velocity within the warehouse was typically below 50 feet per minute, unless a pallet jack or fork lift passed, producing a moderate wind.

The Summer heat stress evaluation was performed on a warm day that was partly cloudy with a peak ambient air (dry bulb) temperature of 93°F. Although this is not the most extreme hot weather one would expect in Columbus, Ohio, it does represent typical conditions of the summer months for this geographic location. The WBGT heat stress measurements recorded spatially throughout the warehouse were reasonably close to one another. Temperatures in the front of the warehouse near the loading dock and numerous bay doors were slightly higher than those in the rest of the building. (The large axial fans in the back of the building exhausted air, which was replaced by outside air infiltrating through the bay doors where trucks were staged.) The early afternoon WBGT measurements ranged from 77.1 - 79.2°F, with a mean of 77.7°F, and the late afternoon average WBGT was 77.0°F. The data obtained from the fixed monitoring location were consistent with these values. The dry bulb and globe temperatures were very similar (demonstrating the lack of a significant radiant source), and the DB increased as the day progressed (from 83.5 to 85.5°F). As the day proceeded into the late afternoon and early evening, the WBGT measurement slowly decreased, largely due to the wet bulb temperature reductions. The data collected on the pallet jacks suggested that any cooling effect obtained from riding the jacks was insignificant, since the workers typically walk the jack to each grocery slot and only ride between aisles and back to the dispatch office.

In order to apply the WBGT measurement to the evaluation criteria, an estimate of the metabolic rate necessary to perform the work and a characterization of the work-rest regimen are required. The heart rate monitoring and oxygen consumption measurements allowed the investigators to identify the specific metabolic rates during the measured order selection. In addition to this estimation, Table 13 presents the metabolic rate estimation based on the job cycle tasks utilizing the metabolic rate approximations provided in Table 12. Both of these methods provided estimates of metabolic rates which were reasonably close to each other, ranging from 5.0 to 7.1 kcal per minute. The metabolic rate of the grocery selector job was determined to be moderate to high, in regards to a heat stress evaluation. The production standards and work shift schedule during the order selecting resulted in a 100% work-0% rest regimen.

Health Hazard Evaluation Report No. 91-405

The NIOSH Recommended Exposure Limit (REL) to environmental heat for heat acclimatized workers functioning at a metabolic rate of 400 kcal/hr and a 100% work-0% rest cycle is a WBGT of 79°F (NIOSH, 1986). The ACGIH Threshold Limit Value (TLV) WBGT for a moderate to high work rate (approximately 400 kcal/hr) and a 100 work-0% rest cycle is also 79°F. The WBGT heat exposure observed in the warehouse during a typical summer day <u>approached these criteria</u>. During more extreme heat conditions of peak summer weather the heat stress exposure for the grocery selectors would probably exceed the evaluation criteria.

Prolonged exposure to excessive heat may cause increased irritability, lassitude, decrease in morale, increased anxiety, and inability to concentrate. The acute physical disabilities caused by excessive heat exposure are, in order of increasing severity: heat rash, heat cramps, heat exhaustion, and heat stroke. Refer to Appendix C for a description of these acute heat stress effects. There are other concerns besides health effects from excessive exposure to heat stress. Ramsey et al describe an increase in unsafe acts associated with exposure to environmental heat, as well as with increased level of physical work (Ramsey et al, 1983).

The control of occupational heat exposure can be accomplished by addressing the heat balance components that significantly contribute to heat gain (stress). The four environmental heat exchange components which potentially contribute to heat stress and possible methods of control are described below (NIOSH, 1986; ACGIH, 1988; Belding, 1973):

<u>Metabolic heat</u> - Metabolic heat can be reduced by mechanization of some or all tasks, increasing rest time, reducing work pace, and sharing the work load with additional workers (particularly during peak heat periods).

<u>Radiant heat gain</u> - Reduction of radiant heat gain can be accomplished by shielding the worker line of sight to the radiant source, insulating radiant heat sources, using reflective screens, wearing radiant reflective clothing (especially if the worker directly faces the source), and covering exposed body parts. None of these apply to Big Bear Warehouse since radiant heat was not a significant contributor to the heat load.

<u>Convective heat exchange</u> - Heat can be gained or lost by convection depending on the air temperature. If the air temperature exceeds the mean skin temperature (considered to be 95°F), then increasing air movement across the skin will contribute to convective heat gain. Control of heat gain under these conditions will require reducing the air temperature, reducing air velocity, and wearing lose fitting (single layer) clothing. If the air temperature is below 95°F, increasing the convective heat loss can be accomplished by increasing air velocity across the skin, removing clothing (maximizing exposed skin surface), and decreasing the air temperature.

<u>Evaporative heat loss</u> - The maximum evaporative cooling capacity of the environment can be expanded by increasing the air velocity and by decreasing the water vapor pressure (humidity) of the work atmosphere. Consideration must be given to the potential of convective heat gain when increasing air velocity, since the benefit of evaporative cooling may be overcome by the convective heat gain due to high air temperatures ($\geq$95°F).

In addition to modifying the work place environmental conditions, the risk of a serious incident due to excessive heat exposure can be reduced by the implementation of a heat stress management program. This is especially important when modification of the environmental conditions is not feasible. The elements of a comprehensive heat stress management program is provided in Appendix D.

Health Hazard Evaluation Report No. 91-405

VI. Conclusions

The objective of an ergonomic job analysis is to fit the job to the worker so that one can work without excessive physical stress, fatigue or harm to one's health. This investigation clearly showed the importance of obtaining quantitative data on biomechanical and physiological responses of workers as they are actually performing their job. The collection of this data has allowed us to make conclusions about the risk of low back injury among grocery selectors in Warehouse 1.

In summary, the order selectors have an elevated risk for musculoskeletal disorders, including low back pain, because of the combination of adverse job factors all contributing to fatigue including heat stress in hot weather, a high metabolic load, and the workers' inability to regulate their work rate because of the work requirements. According to recognized criteria defining worker capability and accompanying risk of low back injury, the job of order selector at this work site will place even a highly selected work force at substantial risk of developing low back injuries. Moreover, in general, we believe that the existing performance standards encourage and contribute to these excessive levels of exertion.

These overall conclusions are based upon the following specific findings from our quantitative evaluation.

1) Based on the revised NIOSH lifting equation, all of the lifting tasks that were sampled exceeded the RWL by significant margins or exceeded maximum limits set for individual task parameters, such as the horizontal reach factor, indicating that the tasks were highly ergonomically stressful.

2) Estimated compressive force on the L5/S1 disc exceeded the recommended biomechanical limit of 770 lbs (3.4 KN), identified by NIOSH as an upper limit for protecting most workers from the risk of low back injury.

3) The results indicate that movements in the sagittal plane (forward bending) required the greatest spinal movement, regardless of the height of the lift. In particular, the low height lifts were associated with the greatest sagittal flexion angle (56 degrees) and highest velocities and accelerations.

4) Based on the fact that spinal forces increase as the flexion angle increases, the results here indicate that 53% of lifts (those included in the low category) would be associated with the greatest biomechanical spine loading, especially when the increased accelerations are considered.

5) Although Marras's model has not been fully validated, it indicates a high probability that warehouse grocery selector tasks would be categorized as "high risk" jobs. This would suggest that it is likely that there would be more than 12 injuries/200,000 hours of work exposure for grocery selectors, a prediction borne out by the company's OSHA logs and workers' claims and the questionnaire survey.

6) Based on average energy demands, the job of order selector exceeded the established criterion of 5 kcal/min (4 METS) for an 8-hr day, which is recognized as moderate to heavy work for a majority of healthy workers.

7) The order selectors at the worksite frequently had an average heart rate of more than 110 beats per minute, which has been suggested as the minimum acceptable for the majority of healthy workers.

8) The order selectors' physiologic energy demands from continuous lifting at a rate of 4.1 lifts/minute (246 lifts/hour) would probably result in fatigued muscles, especially when extended shifts of 10 hours are considered.

9) Ergonomic evaluations showed that the racks were very deep, and the stacking arrangement resulted in most grocery items being located either too high or too low. In addition, the order selectors are required to lift heavy and bulky loads to a vertical height which exceeds the reach limit for most people, and at a horizontal distance which is close to the functional reach limit.

10) Data collected from OSHA 200 logs, workers' compensation logs, and employee questionnaires all found a high rate of back injuries amongst the order selectors.

11) The heat stress exposure during grocery selecting on a moderately hot summer day approached the NIOSH and ACGIH WBGT heat stress criterion. This was largely due to the work activities that resulted in moderate to high metabolic heat production.

12) The dry bulb temperature of the air within the warehouse was cool enough (<95°F) to provide for convective heat loss if air velocity were increased.

13) A source of cool potable water was not easily accessible (on the pallet jacks) at the immediate vicinity of the order selecting.

14) The company lacked a comprehensive heat stress management program, including effective medical examinations and policies.

15) Self-regulation of work activity by selector employees, an important safeguard which reduces the potential for a serious heat related incident, is not compatible with maintaining the incentive standards.

16) Workers in the warehouse reported significantly less control over their work compared to other workers. Much of this lack of control was attributed to the production standards.

VII. Limitations

This investigation performed a number of assessments of the ergonomic characteristics and potential for heat stress of specific job tasks performed by order selectors. Since these investigations were performed during a portion of only three days it is difficult to know whether our assessment is representative of the usual selector workload. In some of these evaluations we focused on job tasks which were considered to be most hazardous although fairly common. As a result, some of our analyses are not necessarily representative of average job tasks. However, the job analyses do provide an accurate description of ergonomic hazards at this warehouse. Because the job tasks are highly repetitive, short sampling periods should yield data representative of usual job tasks. The conclusions of this investigation are supported by the consistency of the results of different exposure assessments and analysis of the health information from the worker compensation records, OSHA logs and the questionnaire survey.

Health Hazard Evaluation Report No. 91-405

VIII. Recommendations

A. Ergonomic

The selectors' current workload requires excessive metabolic and biomechanical demands even for a highly select population. These hazards are not the result of unique characteristics of the work methods or workplace layout, but rather result from a combination of the amount of weight lifted per hour, number of lifts per hour, and lifting of objects from floor level and above waist height. Substantial changes in work organization and methods are needed in order to reduce the hazard. These changes can be accomplished in a variety of ways and should be done though the consultation of a trained ergonomist. Some of the specific recommendations are:

1) Ergonomic principles should be used in the design of racks, physical layout, size of the order, and arrangement of grocery items. This should minimize the physiological cost of work, reduce injury and illness to order selectors, and may even improve productivity.

2) Jobs should be analyzed using the NIOSH lifting equation to identify highly stressful tasks and evaluate alternate methods or workplace layouts.

3) Since both the physical strength and endurance requirements for the order selector's job are very high, even with the recommended changes, certain administrative will be necessary, such as worker rotation, to reduce the future risk of injury and illness among the order selectors.

4) Performance standards, if they are to be used, must be based on objective measures of physical effort, such as heart rate and oxygen uptake, in order to determine reasonable workloads which will not place the worker at an excess risk of injury.

5) Light duty jobs should be made available for injured workers in order to encourage and facilitate their return to work.

6) Overtime should be kept to a minimum, as the energy requirements for the job are very high. Further, overtime should be made voluntary so that a tired worker with a limited aerobic capacity is not forced to maintain a certain level of performance after an eight-hour work day.

7) Warehouses should develop a better system to monitor injury and illness profiles of the workers. This system can be useful in preparing injury and illness statistics and monitoring the effectiveness of workplace intervention programs.

8) The heaviest objects should be stored in a walk in slots with the bottom raised to knee height. Less heavy items should be stored in the bottom slot of a two slot configuration, rather than in the top slot and only light and non-breakable items should be stored in the top racks.

9) The size of an order should be restricted so that the pallet load heights do not exceed 60 inches.

10) Using single pallet rather than double pallet orders is one method to decrease the cycle time and increase the frequency of recovery periods between orders.

Health Hazard Evaluation Report No. 91-405

11) The use of computer-generated orders should be modified to ensure that order selections for each worker is balanced with respect to the "load difficulty." A heavy or difficult load should be followed by a less difficult load to allow the worker an opportunity for some recovery. Factors which appear to affect load difficulty were well-known to the workers including: weight, number of items, location, and type of items to be selected.

B. Heat Stress

A number of control options are available for addressing occupational heat exposure; these include engineering controls, administrative controls, and personal protective equipment. Typically, implementation of a single control will not adequately address the entire heat stress problem. It is prudent to identify the most significant conditions that contribute to the heat stress so that the initial focus of the corrective measures address the most hazardous heat balance component. During hot periods, when worker exposure exceeds or approaches the WBGT criterion, the following recommendations should be considered:

1) The grocery selectors' metabolic heat production should be reduced by relaxing the work pace and/or by increasing rest time in cool locations.

2) Provide space cooling fans in the aisles of the warehouse to maximize evaporative cooling due to the increased air velocity.

3) When exterior temperatures are significantly higher than those in the warehouse, uncooled outside air should not be brought into the building by operating the (existing) exhaust fans. (Implementation of this recommendation requires alternative equipment to create air movement and circulation within the warehouse; otherwise, failure to operate the exhaust fans would reduce the air movement and adversely impact evaporative heat loss.)

4) Install ceiling vents to assist in removing heat from within the building.

5) Provide large-capacity supply air fans on the dock side of the warehouse (on the opposite side of the warehouse from the existing exhaust fans already present). These fans should be operated in conjunction with the exhaust fans in the evenings and night hours to bring cool air into the warehouse and remove hot air.

6) Ensure that extra precautionary measures are implemented for protecting unacclimatized workers such as reduced work rate expectations, increased resting time, and gradually increased work loads to allow for safe heat acclimatization. This is especially important during sudden heat waves or when a worker returns to work after a prolonged absence.

7) Provide cool drinking water on each pallet jack so that order selectors have easy accessibility to water at their work locations.

8) Implement a comprehensive heat management program, which includes heat alert policies that are based on environmental conditions.

IX. References

ACGIH [1988]. *Industrial Ventilation: A manual of recommended practice, 20th ed.* Cincinnati OH: American Conference of Governmental Industrial Hygienists.

ACGIH [1990]. *Threshold limit values and biological exposure indices for 1990-1991*. Cincinnati, OH: American Conference of Governmental Industrial Hygienists.

AIHA [1975]. *Heat exchange and human tolerance limits, heating and cooling for man in industry, 2nd ed.* Akron, OH: American Industrial Hygiene Association. pg 5-28.

Andersson, G., [1981]. Epidemiologic aspects on low-back pain in industry, *Spine*, 6(1): 53-60.

Astrand, P. and Rodahl, K., [1986]. *Textbook of Work Physiology: Physiological Bases of Exercise*, McGraw-Hill Book Company.

Ballal, MA. and Macdonald, IA., [1982]. An evaluation of the oxylog as a portable device with which to measure oxygen consumption, *Clin. Phys. Physiol. Meas.*, 3(1):57-65.

Bean, JC, Chaffin, DB, and Schultz, AB, [1988]. A biomechanical model calculation of muscle contraction forces: A double linear programming method, *Journal of Biomechanics*, 21(1):59-66.

Belding H, Hatch T [1955]. Index for evaluating heat stress in terms of resulting physiological strain. *Heat Pip Air Condit* 27:129.

Belding H [1973]. Control of exposures to heat and cold, Chapter 38 In: *The Industrial environment - its evaluation and control.* Cincinnati OH: U.S. Department of Health, Education, and Welfare, Public Health Service, Health Services and Mental Health Administration, National Institute for Occupational Safety and Health; DHEW (NIOSH) contract no. HSM-99-71-45.

Bigos, SJ, Spengler, DM, Martin, NA, Zeh, J, Fisher, L, Nachemson, A, and Wang, MH, [1986]. Back injuries in industry: A retrospective study II. Injury factors," *Spine*, 11(3).

Borg, GA, [1982]. Psychological bases of perceived exertion, *Medicine and Science in Sports and Exercise*, 14(5):377-381.

Borg, GA., [1990]. Psychological scaling with applications in physical work and the perception of exertion," *Scand J Work Environ Health*, 16(suppl 1):55-58.

BLS [1992]. Employment and Earnings. Washington D.C.: U.S. Department of Labor, Bureau of Labor Statistics.

Caplan, RD, Cobb, S, French, JR, Harrison, RV, and Pinneau [1975]. Job demands and worker health, *HEW Publication No. (NIOSH) 75-160.*

Chaffin, DB, Herrin, GD, Keyserling, WM, and Garg, A., [1977]. A method for evaluating the biomechanical stresses resulting from manual materials handling jobs, *American Industrial Hygiene Association Journal*, 38:662-675.

Chaffin, DB and Andersson, GBJ., [1991]. *Occupational Biomechanics, second edition*, John Wiley & Sons, Inc.

Chaffin, DB and Ayoub, MN, [1975]. The problem of manual materials handling, *Industrial Engineering*, July:24-29.

Corlett EN and Bishop RP, [1976]. A technique for assessing postural discomfort. Ergonomics 19(2):1975-1982.

Dukes-Dobos FN [1981]. Hazards of heat exposure: A review. *Scand J Work Environ Health* 7:73-83.

Eastman Kodak Company, [1986]. *Ergonomic Design for People at Work, vol. 2.*

Garg, A [1986]. Biomechanical and ergonomic stresses in warehouse operations, " *IIE Transactions,* Sept.: 246-250.

Garg, A and Badger, D, [1986]. Maximum acceptable weights and maximum voluntary isometric strengths for asymmetric lifting, *Ergonomics,* 29(7): 879-892.

Garg, A, Chaffin, DB, and Herrin, GD, [1978]. Prediction of metabolic Rates for manual materials handling jobs," *American Industrial Hygiene Association Journal,* 39(8): 661-674.

Guo, H, Tanaka, S, Cameron, LL, Seligman, PJ, Behrens, VJ, Ger, J., Wild, DK, and Putz-Anderson, V, [1993]. Back pain among workers in the United States: National estimates and workers at high risk, NIOSH, *Presentation at the Epidemiologic Intelligence Service Conference, CDC, April 1993.*

Greenberger, DB [1981]. Personal control at work: Its conceptualization and measurement. *Technical Report 1-1-14,* University of Wisconsin-Madison.

Herrin, GD, Jaraiedi, M, and Anderson, CK, [1986]. Prediction of overexertion injuries using biomechanical and psychophysical models, *American Industrial Hygiene Association Journal,* 47(6): 322-330.

Houghton F, Yaglou C [1923]. Determining lines of equal comfort. *J Am Soc Heat and Vent Engrs* 29:165-176.

Hurrell JJ, Linstrom K, [1992]. Comparison of job demands, control and psychosomatic complaints at different career stages of manager in Finland and the United States. *Scand J Work Environ Hlth,* 18(suppl 2):11-13.

ISO [1982]. *Hot environments - estimation of heat stress on working man based on the WBGT index.* Geneva, Switzerland: International Standards Organization. ISO No. 7243-1982.

Klein, BP, Jensen, RC, and Sanderson, LM, [1984]. Assessment of worker's compensation claims for back strains/sprains, *Journal of Occupational Medicine,* 26(6):443.

Kourinka I, Jonsson B, Kilbom A, Vinterberg H, Biering-Sorensen F, Andersson G, et al. [1987]. Standardized Nordic questionnaires for the analysis of musculoskeletal symptoms. *Appl Ergonomics,* 18(3):233-237.

Landsbergis PA, [1988]. Occupational stress among health care workers: A test of the job demands-control model. *Journal of Organizational Behavior,* 9:217-239.

Magora, A., [1973]. Investigation of the relation between low back pain and occupation, *Scandanavian Journal of Rehab. Med.,* 5:186-190.

Marras, WS, Lavender, SA, Leurgans, SE, Rajulu, SL, Allread, WG, Fathallah, FA, and Ferguson, SA, [1993]. The role of dynamic three-dimensional trunk motion in occupationally-related low back disorders: The effects of workplace factors, trunk position, and trunk motion characteristics on risk of injury, *Spine,* 18(5):617-628.

McLaney, MA and Hurrell, JJ. [1988]. Control, stress and job satisfaction in Canadian nurses, *Work & Stress*, 2(3):217-224.

NAWGA and IFDA, [1992]. *Voluntary ergonomics program management guidelines*, Publication of the National American Wholesale Grocers' Association and International Foodservice Distributors Association.

NIOSH [1972]. *Criteria for a recommended standard: occupational exposure to hot environments.* Cincinnati, OH: U.S. Department of Health, Education and Welfare, Health Services and Mental Health Administration, National Institute for Occupational Safety and Health, DHEW(NIOSH) Publication No. 72-10269.

NIOSH [1981]. *Work practices guide for manual material handling.* Cincinnati, OH: U.S. Department of Health and Human Services, Public Health Service, Centers for Disease Control, National Institute for Occupational Safety and Health. DHHS(NIOSH) Publication No. 81-122.

NIOSH [1986]. *Criteria for a recommended standard: occupational exposure to hot environments, revised criteria.* Cincinnati, OH: U.S. Department of Health and Human Services, Public Health Service, Centers for Disease Control, National Institute for Occupational Safety and Health. DHHS(NIOSH) Publication No. 86-113.

Ramsey J, Burford C, Beshir M, Jensen R [1983]. Effects of workplace thermal conditions on safe work behavior. *J Safety Research* 14:105-114.

Silverstein BA, Fine LJ, Armstrong TJ [1986]. Hand wrist cumulative trauma disorders in industry *Br J Ind Med*, 43:779-784.

Snook SH, Campanelli RA, and Hart JW [1978]. A study of three preventive approaches to low back injury, *Journal of Occupational Medicine*, 20(7):478-481.

U.S. Department of Defense [1980]. *Occupational and environmental health: prevention, treatment and control of heat injuries.* Washington, DC: Headquarters, Departments of the Army, Navy, and Air Force, TB MED 507, NAVMED P-5052-5, AFP 160-1.

U. S. Department of Labor, Bureau of Labor Statistics [1992]. Annual Survey of Occupational Injuries and Illnesses in the U. S., 1990; Bulletin #2399.

Viikari-Juntura E [1983]. Neck and upper limb disorders among slaughterhouse workers. *Scand J Work and Environ Hlth*, 9:283-290.

Waters TR, Putz-Anderson V, Garg A, Fine L [1993]. Revised lifting equation for the design and evaluation of manual lifting tasks. *Ergonomics*, 36:749-776.

Waters, TR, [1993]. Workplace factors and trunk motion in grocery selector tasks. *to be presented at Human Factors and Ergonomic Society Annual Meeting, October 1993.*

Webster, BS and Snook, SH, [1990]. The cost of compensable low back pain, *Journal of Occupational Medicine*, 32:13-15.

WHO [1969]. Health factors involved in working under conditions of heat stress. *Geneva, Switzerland: World Health Organization.* Technical Report Series No. 412.

Yaglou C, Minard D [1957]. Control of heat casualties at military training centers. *Arch Indust Health* 16:302-316.

Health Hazard Evaluation Report No. 91-405

X. Authorship and Acknowledgements

Principal Investigators:
Vernon Putz-Anderson, PhD
Section Chief
Psychophysiology and Biomechanics Section

Thomas Waters, PhD
Research Physiologist
Psychophysiology and Biomechanics Section
Applied Psychology and Ergonomics Branch
Division of Biomedical and Behavioral Science

Sherry Baron, MD, MPH
Medical Officer
Medical Section

Kevin Hanley, MSPH, CIH
Industrial Hygienist
Industrial Hygiene Section
Hazard Evaluations and Technical
 Assistance Branch
Division of Surveillance, Hazard
 Evaluations, and Field Studies

Field Assistants:
Glenn Doyle, BS
Psychological Technician

Libby Steward, MA
Research Psychologist
Psychophysiology and Biomechanics Section
Applied Psychology and Ergonomics Branch
Division of Biomedical and Behavioral Science

Jonathan Dropkin, PT
Health Scientist
Medical Section
Hazard Evaluations and Technical
 Assistance Branch
Division of Surveillance, Hazard
 Evaluations, and Field Studies

Originating Office:
Hazard Evaluations and Technical
 Assistance Branch
Division of Surveillance, Hazard
 Evaluations, and Field Studies

Health Hazard Evaluation Report No. 91-405

XI. DISTRIBUTION AND AVAILABILITY

Copies of this report may be freely reproduced and are not copyrighted.

Copies of this report have been sent to:

1. Big Bear Stores Company, Columbus, Ohio
2. United Industrial Workers Union
3. Confidential Requestors

For the purpose of informing affected employees, copies of this report shall be posted by the employer in a prominent place accessible to the employees for a period of 30 calendar days.

Table 1
OSHA 200 Log Entries for Selectors
Big Bear Grocery Warehouse 1

Year	All Injuries		Back Injuries		Back-Related Lost Workdays	
	# Cases	Rate	# Cases	Rate	Total #	Average #[1]
1991	35	52%	19	28%	642	34
1990	54	81%	21	31%	932	44
1989	58	87%	26	39%	468	18
1988	59	88%	21	31%	710	34
1987	53	79%	19	28%	288	15

[1] Average number of lost-workdays for each back injury case

Table 2
Demographics of Study Participants
Big Bear Grocery Warehouse 1

	Full-Time Selectors	Part-Time Selectors
Average Age (range)	30 yrs (19-47)	25 yrs (19-31)
% Male	100%	100%
Average Yrs at Big Bear (Range)	4 yrs (1-12)	less than 1 yr
Average Yrs at Other Warehouse (range)	3 yrs (0-18)	less than 1 yr (0-5)
Average Height (range)	71 inches (66-76)	71 inches (66-76)
Average Weight (Range)	189 lbs (150-290)	192 lbs (158-233)

Table 3
Injuries and Missed Workdays Reported by Questionnaire
Occurring During the Previous 12 Months
Big Bear Grocery Warehouse 1 Full-Time Selectors Only

Body Part	Injuries		Average # Missed Workdays	Range of Missed Workdays
	Number[1]	Rate		
Back	6	18%	91	0-510
Neck/Shoulder	5	13%	26	0-90
Lower Extremity	5	13%	2	0-5
Hand	3	8%	3	0-7

[1] This column adds up to 20 because one worker reported two injuries (Back and Hand)

Table 4
Rate of Reported Discomfort
Big Bear Grocery Warehouse 1 Selectors

Body Part	Full-Time Selectors (N = 38)	Part-Time Selectors (N = 10)
Neck/Shoulder	37%	60%
Back	71%	40%
Lower Extremity	37%	10%
Hand/Arm	24%	40%

Table 5
Summary results of NIOSH Lifting Equation Evaluation
Big Bear Grocery Warehouse 1 Selectors

TASK #	LOAD	LIFTING EQUATION COMPONENTS							RESULTS	
		LC	HM	VM	DM	CM	AM	FM	RWL	LI
1	30 lbs	51	0.46	0.81	0.89	0.95	1.00	0.45	7.2 lbs	4.2
2	38 lbs	51	0.40*	0.87	1.00	0.95	0.89	0.45	6.8 lbs	5.6
3	42 lbs	51	0.40*	0.92	0.89	0.95	0.73	0.45	5.2 lbs	8.1
4	38 lbs	51	0.40*	0.78	1.00	1.00	0.72	0.45	5.2 lbs	7.3
5	58 lbs	51	0.48	0.89	1.00	0.95	0.77	0.45	7.2 lbs	8.0

* Actual horizontal distances exceeded 25 inches, which according to the NIOSH equation would set HM equal to 0.0, resulting in RWL of 0 and the requirement to redesign the job. For this exercise, the HM was set to the maximum of 25 inches for comparison purposes.

TABLE 6
Summary Results of 3-D Biomechanical Analysis
Big Bear Grocery Warehouse 1 Selectors

Task	Load	Disc Comp.	% Capable*		Shear Force	Torsion
			Hip	Torso		
1	30 lbs	930 lbs	63	94	123 lbs	69 ft-lb
2	38 lbs	830 lbs	70	95	131 lbs	113 ft-lb
3	42 lbs	896 lbs	16	76	116 lbs	40 ft-lb
4	38 lbs	662 lbs	52	59	87 lbs	61 ft-lb
5	58 lbs	801 lbs	55	86	112 lbs	68 ft-lb

* The minimum percentage of the male working population who have sufficient strength for that lifting task

Table 7
Lifting Task Conditions by Origin and Destination
Big Bear Grocery Warehouse 1 Selectors

Origin	Destination			
	Low	Medium	High	Total
Low	15 (7%)	36 (17%)	29 (13%)	80 (37%)
Medium	10 (5%)	16 (7%)	34 (16%)	60 (28%)
High	24 (11%)	28 (13%)	24 (11%)	76 (35%)
Total	49 (23%)	80 (37%)	87 (40%)	216 (100%)

Table 8
Dynamic Lifting Task Analysis
Big Bear Grocery Warehouse 1 Selectors

Height Range	Lift Rate (Lifts/HR)	Max. Moment (N·M)	Maximum Sagittal Flexion (Degrees)	Average Transverse Velocity (Deg/Sec)	Maximum Lateral Velocity (Deg/Sec)
Low	246	92.0	55.7	5.9	31.0
Medium	246	126.0	37.5	5.8	33.3
High	246	87.0	28.0	6.5	26.0

Table 9
Summary Table For Metabolic Criteria
Big Bear Grocery Warehouse 1 Selectors

Participant	Variables	Value
1	Total Cases/order	167
	Total Weight/order	2198 lbs
	Allowed Time/order	34.9 min
	Weight/min	63 lbs/min
	Performance Index*	116%
	Working Metabolic Rate	5.4 kcal/min
	Working Heart Rate	122 beats/min
	Predicted Metabolic Rate	6.0 kcal/min
2	Total Cases/order	138
	Total Weight/order	4220 lbs
	Allowed Time/order	36.7 min
	Weight/min	115 lbs/min
	Performance Index	116%
	Working Metabolic Rate	5.9 kcal/min
	Working Heart Rate	104 beats/min
	Predicted Metabolic Rate	5.0 kcal/min
3	Total Cases/order	101
	Total Weight/order	3862 lbs
	Allowed Time/order	25.8 min
	Weight/min	150 lbs/min
	Performance Index	143%
	Working Metabolic Rate	8.0 kcal/min
	Working Heart Rate	131 beats/min
	Predicted Metabolic Rate	7.6 kcal/min

* Performance Index = (allowed time per order/actual time per order) * 100

Table 10
Heat and Humidity Measures on May 14, 1992
Big Bear Grocery Warehouse 1

GRID, EARLY PM, (13:05 - 14:40)					
	WB*	DB*	GT*	WBGT*	RH
Minimum	57.2	73.1	74.3	62.3	
Maximum	60.3	75.9	77.5	65.4	
Mean	58.4	74.4	75.5	63.5	37.5%
Outside (shade)	55.1	69.6	76.2	61.4	37.5%

GRID, LATE PM, (17:00 - 18:15)					
	WB	DB	GT	WBGT	RH
Minimum	56.7	73.6	75.0	62.3	
Maximum	59.3	77.4	77.5	64.7	
Mean	58.1	75.2	75.7	63.4	35%
Outside - shade, (18:50)	49.0	62.3	64.6	53.7	37.5%

STATIONARY, (12:25 - 18:30)					
	WB	DB	GT	WBGT	RH
Minimum	57.9	73.6	74.7	63.0	
Maximum	60.5	76.3	77.1	65.5	
Mean	59.1	75.2	76.2	64.2	37.5%

* Degrees Fahrenheit

Table 11
Heat and Humidity Measures on July 14, 1992
Big Bear Grocery Warehouse 1

GRID, EARLY PM, (11:35 - 13:05)	WB°	DB°	GT°	WBGT°	RH
Minimum	74.4	82.6	83.5	77.1	
Maximum	76.0	87.4	85.3	79.2	
Mean	74.9	84.1	84.5	77.7	65%
Outside (shade)	75.5	92.8	100.6	83.0	45%

GRID, LATE PM, (16:50 - 18:15)	WB	DB	GT	WBGT	RH
Minimum	72.9	84.2	84.9	76.5	
Maximum	73.9	88.7	88.8	77.9	
Mean	73.3	85.4	85.8	77.0	55%
Outside (shade)					
16:50	73.4	89.9	93.8	79.5	45%
18:15	73.0	88.6	90.2	78.1	47.5%

STATIONARY, (11:15 - 17:50)	WB	DB	GT	WBGT	RH
Minimum	73.1	83.5	83.2	76.7	
Maximum	75.1	85.5	85.6	77.8	
Mean	74.0	84.6	84.8	77.2	60%

° Degrees Fahrenheit

Table 12
Metabolic Heat Production Rates by Task Analysis
Big Bear Grocery Warehouse 1

A. Body Position and Movement		kcal/min*
Sitting		0.03
Standing		0.6
Walking		2-0-3.0
Walking uphill		add 0.8 per meter rise
B. Type of Work	**Average kcal/min**	**Range kcal/min**
Hand work		
light	0.4	0.2-1.2
heavy	0.9	
Work one arm		
light	1.0	0.7-2.5
heavy	1.8	
Work both arms		
light	1.5	1.0-3.5
heavy	2.5	
Work whole body		
light	3.5	2.5-9.0
moderate	5.0	
heavy	7.0	
very heavy	9.0	
C. Basal metabolism	1.0	
D. Sample calculation**		**Average kcal/min**
Assembling work with heavy hand tools		
1. Standing		0.6
2. Two-arm work		3.5
3. Basal metabolism		1.0
Total		5.1 kcal/min

* For standard worker of 70 kg body weight (154 lbs.) and 1.8 m^2 body surface (19.4 ft^2).
** Example of measuring metabolic heat production of a worker when performing initial screening.

Table 13.
Estimated Metabolic Rate, Grocery Selectors
Big Bear Grocery Warehouse 1

	Range[1] (Kcal/min)	Estimate (Kcal/min)
Body Position 1. Always standing and walking	0.6 2.0-3.0	1.5

Type of Work[2]
 	
 1. LIFTING and CARRYING
  Whole body - moderate weights, repetitive lifts, high pace.
 	
 2. PALLET FORMING
  Both Arms - with bending and reaching
	6.0
	
	 2.0
		
		 4.5[2]

Basal Metabolism	1.0	1.0
Summation		7.0
Hourly Estimation		420
Metabolic Rate Work Category		Moderate to high

NOTES:
[1] From Reference 1, Table 12.
[2] Worker spends most of the work cycle lifting and carrying (6.0 Kcal/min), with the remainder of time spent pallet forming (2.0 Kcal/min) and walking pallet jack to the next slot.
[3] kcal/min = kilocalories per minute.

Task 1 - Sweet Peas

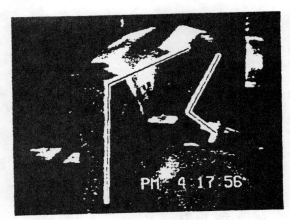
Task 2 - Gatorade

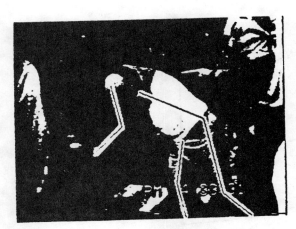
Task 3 - Pet Club Variety Mix

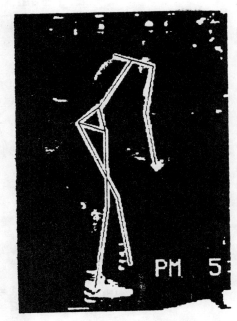
Task 4 - Miracle Whip

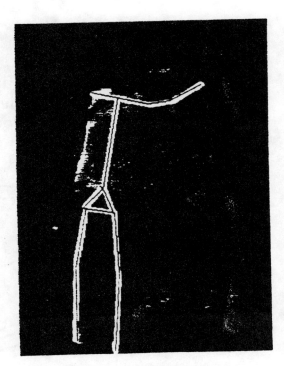
Task 5 - Distilled Water

Figure 1. Examples of lifting tasks analyzed using revised NIOSH Equation. Big Bear Grocery Warehouse 1

Figure 2
Body Map Discomfort Diagram
Big Bear Grocery Warehouse 1

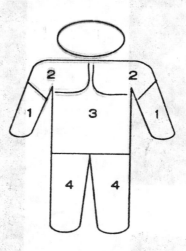

1 = Hand/Arm
2 = Neck/Shoulder
3 = Back
4 = Lower Extremity

Figure 3
NIOSH Recommended Exposure Limits
Big Bear Grocery Warehouse 1

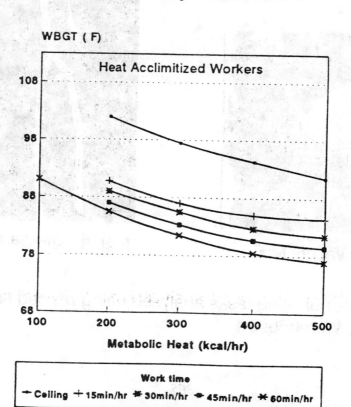

Appendix A
Worker Questionnaire
Big Bear Grocery Warehouse 1

1. What is your name:(LAST)_____(FIRST)_____

2. What is your: age_____yrs,

   height_____ft_____in,

   weight_____lbs

3. When did you start working at Big Bear Warehouse?

   _____month,19_____yr

4. Total years worked at Big Bear Warehouse is:_____yr(s)

5. Did you work at another warehouse previously?_____yes,_____no

6. If yes, how many years did you work at that other warehouse?

   _____yrs,_____months

7. During the past year (July 1, 1991--June 30 1992), have you ever had an injury at work:_____yes,_____no

8. If yes, what part of your body did you injure ?_____

9. During the past year, have you ever missed any workdays due to an injury at work ?_____yes,_____no

10. If yes, how many days did you miss ?_____days

11. What is your telephone number:(   )_____

12. Using the rating scale shown below please rate the OVERALL physical effort level demanded by your job today?

   Please circle the most appropriate number on the following scale

   20
   19 - Very, very hard
   18
   17 - Very hard
   16
   15 - Hard
   14
   13 - Somewhat hard
   12
   11 - Fairly light
   10
   9 - Very light
   8
   7 - Very, very light
   6

13. Have you had any pain or discomfort during the last year?   1. YES   2. NO   Go to Next page.

14. If YES, put a number in EACH box to indicate your level of discomfort, using the following scale.

   0 -No discomfort 1- Uncomfortable 2 -Very uncomfortable 3 -Extremely uncomfortable

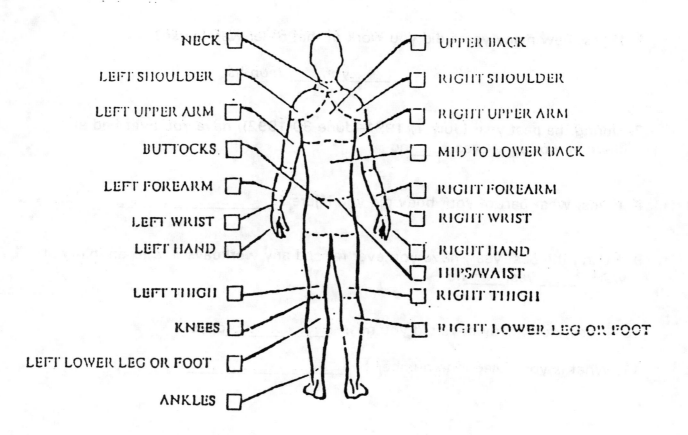

15. Please (circle) the relevant number next to the question.

How often:	Rarely	Occasionally	Sometimes	Fairly Often	Very Often
Does your job require you to work very fast?	1	2	3	4	5
Does your job require you to work very hard?	1	2	3	4	5
Does your job leave you with little time to get things done?	1	2	3	4	5
Is there a great deal to be done?	1	2	3	4	5
Is there a marked increase in the workload?	1	2	3	4	5
Is there a marked increase in the amount of concentration required on your job?	1	2	3	4	5

16. Please (circle) one number for each word which describes how you have been feeling during the past week, including today.

	Not At ALL	A Little	Moderately	Quite A Bit	Extremely
Worn Out	1	2	3	4	5
Energetic	1	2	3	4	5
Fatigued	1	2	3	4	5

17. Please (circle) the appropriate number next to the statement.

	Strongly Disagree				Strongly Agree
There is constant work pressure to keep working.	1	2	3	4	5
There always seems to be an urgency about everything.	1	2	3	4	5
People cannot afford to relax.	1	2	3	4	5
Nobody works too hard.	1	2	3	4	5
There is no time pressure.	1	2	3	4	5
It is very hard to keep up with your work load.	1	2	3	4	5
You can take it easy & still get your work done.	1	2	3	4	5
There are always deadlines to be met.	1	2	3	4	5

18. The next series of questions ask how much influence or control you have at work. Please (circle) the appropriate number corresponding to the question.

How much influence do you have over the:	Very Little	Little	Moderate	Much	Very Much
Variety of tasks you perform?	1	2	3	4	5
Amount of work you do?	1	2	3	4	5
Order in which you perform tasks at work?	1	2	3	4	5
Pace of your work, that is how fast or slow do you work?	1	2	3	4	5
Quality of the work that you do?	1	2	3	4	5

	Very Little	Little	Moderate	Much	Very Much
To what extent can you do your work ahead & take a short rest break during work hours?	1	2	3	4	5
In general, how much influence do you have over work & work related factors?	1	2	3	4	5

Appendix B
NIOSH Lifting Equation Calculations
Big Bear Grocery Warehouse

A. Calculation for Recommended Weight Limit

RWL = LC * HM * VM * DM * AM * FM * CM
(* indicates multiplication.)

Recommended Weight Limit

Component	METRIC	U.S. CUSTOMARY
LC = Load Constant	23 kg	51 lbs
HM = Horizontal Multiplier	(25/H)	(10/H)
VM = Vertical Multiplier	(1-(.003$\|$V-75$\|$))	(1-(.0075$\|$V-30$\|$))
DM = Distance Multiplier	(.82 + (4.5/D))	(.82 + (1.8/D))
AM = Asymmetric Multiplier	(1-(.0032A))	(1-(.0032A))
FM = Frequency Multiplier	(from Table 1)	
CM = Coupling Multiplier	(from Table 2)	

Where:

H = Horizontal location of hands from midpoint between the ankles. Measure at the origin and the destination of the lift (cm or in).

V = Vertical location of the hands from the floor. Measure at the origin and destination of the lift (cm or in).

D = Vertical travel distance between the origin and the destination of the lift (cm or in).

A = Angle of asymmetry - angular displacement of the load from the sagittal plane. Measure at the origin and destination of the lift (degrees).

F = Average frequency rate of lifting measured in lifts/min.
Duration is defined to be: ≤ 1 hour; ≤ 2 hours; or ≤ 8 hours assuming appropriate recovery allowances (See Table X).

Appendix B
Table 1
Frequency Multiplier (FM)
NIOSH Lifting Equation

Frequency Lifts/min	Work Duration					
	≤ 1 Hour		≤ 2 Hours		≤ 8 Hours	
	V < 75	V ≥ 75	V < 75	V ≥ 75	V < 75	V ≥ 75
0.2	1.00	1.00	.95	.95	.85	.85
0.5	.97	.97	.92	.92	.81	.81
1	.94	.94	.88	.88	.75	.75
2	.91	.91	.84	.84	.65	.65
3	.88	.88	.79	.79	.55	.55
4	.84	.84	.72	.72	.45	.45
5	.80	.80	.60	.60	.35	.35
6	.75	.75	.50	.50	.27	.27
7	.70	.70	.42	.42	.22	.22
8	.60	.60	.35	.35	.18	.18
9	.52	.52	.30	.30	.00	.15
10	.45	.45	.26	.26	.00	.13
11	.41	.41	.00	.23	.00	.00
12	.37	.37	.00	.21	.00	.00
13	.00	.34	.00	.00	.00	.00
14	.00	.31	.00	.00	.00	.00
15	.00	.28	.00	.00	.00	.00
>15	.00	.00	.00	.00	.00	.00

†Values of V are in cm; 75 cm = 30 in.

Appendix B
Table 2
Coupling Multiplier
NIOSH Lifting Equation

Couplings	V < 75 cm (30 in)	V ≥ 75 cm (30 in)
	Coupling Multipliers	
Good	1.00	1.00
Fair	0.95	1.00
Poor	0.90	0.90

Appendix C
Acute Physical Effects Caused by Excessive Heat Stress.
Big Bear Grocery Warehouse

Heat rash - Heat rash (prickly heat) may be caused by unrelieved exposure to hot and humid air. The openings of the sweat ducts become plugged due to the swelling of the moist keratin layer of the skin which leads to inflammation of the glands. There are tiny red vesicles (fluid filled bumps) visible in the affected area and, if the affected area is extensive, sweating can be substantially impaired. This may result not only in discomfort, but in a decreased capacity to tolerate heat.

Heat cramps - Heat cramps may occur after prolonged exposure to heat with profuse perspiration and inadequate replacement of salt. The signs and symptoms consist of spasm and pain in the muscles of the abdomen and extremities, especially in the muscles which are working the hardest. Albuminuria (protein in the urine) may be a transient finding.

Heat exhaustion - Heat exhaustion may result from physical exertion in a hot environment when vasomotor control (regulation of muscle tone in the blood vessel walls) and cardiac output are inadequate to meet the increased demand placed upon them by peripheral vasodilation or the reduction in plasma volume due to dehydration. Signs and symptoms of heat exhaustion may include pallor, lassitude, dizziness, syncope, profuse sweating, and cool moist skin. There may or may not be mild hyperthermia.

Heat stroke - Heat stroke is a medical emergency. An important predisposing factor is excessive physical exertion. Signs and symptoms may include dizziness, nausea, severe headache, hot dry skin due to cessation of sweating, very high body temperature [usually 106 °F (41 °C) or higher], confusion, delirium, collapse, and coma. Often circulation is compromised to the point of shock. If steps are not taken to begin cooling the body immediately, irreversible damage to the internal organs and death may ensue.

Appendix D
Elements of a Comprehensive Heat Stress Management Program.
Big Bear Grocery Warehouse

Written program - A detailed written document is necessary to specifically describe the company procedures and policies in regards to heat management. The input from management, technical experts, physician(s), labor union, <u>and</u> the affected employees should be considered when developing the heat management program. This program can only be effective with the full support of plant management.

Environmental monitoring - In order to determine which employees should be included in the heat management program, monitoring the environmental conditions is essential. Environmental monitoring also allows one to determine the severity of the heat stress potential during normal operations and during heat alert periods.

Medical examinations and policies - Preplacement and periodic medical examinations should be provided to all employees included in the heat management program where the work load is heavy or the environmental exposures are extreme. Periodic exams should be conducted at least annually, ideally immediately prior to the hot season (if applicable). The examination should include a comprehensive work and medical history with special emphasis on any suspected previous heat illness or intolerance. Organ systems of particular concern include the skin, liver, kidney, nervous, respiratory, and circulatory systems. Written medical policies should be established which clearly describe specific predisposing conditions that cause the employee to be at higher risk of a heat stress disorder, and the limitations and/or protective measures implemented in such cases.

Work schedule modifications - The work-rest regime can be altered to reduce the heat stress potential. Shortening the duration of work in the heat exposure area and utilizing more frequent rest periods reduces heat stress by decreasing the metabolic heat production and by providing additional recovery time for excessive body heat to dissipate. Naturally, rest periods should be spent in cool locations (preferably air conditioned spaces) with sufficient air movement for the most effective cooling. Allowing the worker to self-limit their exposure on the basis of signs and symptoms of heat strain is especially protective since the worker is usually capable of determining their individual tolerance to heat. However, there is a danger that under certain conditions, a worker may not exercise proper judgement and experience a heat-induced illness or accident.

Acclimatization - Acclimatization refers to a series of physiological and psychological adjustments that occur which allow one to have increased heat tolerance after continued and prolonged exposure to hot environmental conditions. Special attention must be given when administering work schedules during the beginning of the heat season, after long weekends or vacations, for new or temporary employees, or for those workers who may otherwise be unacclimatized because of their increased risk of a heat-induced accident or illness. These employees should have reduced work loads (and heat exposure durations) which are gradually increased until acclimatization has been achieved (usually within 4 or 5 days).

Clothing - Clothing can be used to control heat stress. Workers should wear clothing which permits maximum evaporation of perspiration, and a minimum of perspiration run-off which does not provide heat loss, (although it still depletes the body of salt and water). For extreme conditions, the use of personal protective clothing such as a radiant reflective clothing, and torso cooling vests should be considered.

Buddy system - No worker should be allowed to work in designated hot areas without another person present. A buddy system allows workers to observe fellow workers during their normal job duties for early signs and symptoms of heat intolerance such as weakness, unsteady gait, irritability, disorientation, skin color changes, or general malaise, and would provide a quicker response to a

Appendix D (cont.)
Elements of a Comprehensive Heat Stress Management Program.
Big Bear Grocery Warehouse

heat-induced incident.

Drinking water - An adequate amount of cool (50-60°F) potable water should be supplied within the immediate vicinity of the heat exposure area as well as the resting location(s). Workers who are exposed to hot environments are encouraged to drink a cup (approximately 5-7 ounces) every 15-20 minutes even in the absence of thirst.

Posting - Dangerous heat stress areas (especially those requiring the use of personal protective clothing or equipment) should be posted in readily visible locations along the perimeter entrances. The information on the warning sign should include the hazardous effects of heat stress, the required protective gear for entry, and the emergency measures for addressing a heat disorder.

Heat alert policies - A heat alert policy should be implemented which may impose restrictions on exposure durations (or otherwise control heat exposure) when the National Weather Service forecasts that a heat wave is likely to occur. A heat wave is indicated when daily maximum temperature exceeds 95°F or when the daily maximum temperature exceeds 90°F and is at least 9°F more than the maximum reached on the preceding days.

Emergency contingency procedures - Well planned contingency procedures should be established in writing and followed during times of a heat stress emergency. These procedures should address initial rescue efforts, first aid procedures, victim transport, medical facility/service arrangements, and emergency contacts. Specific individuals (and alternatives) should be assigned a function within the scope of the contingency plan. Everyone involved must memorize their role and responsibilities since response time is critical during a heat stress emergency.

Employee education and training - All employees included in the heat management program or emergency contingency procedures should receive periodic training regarding the hazards of heat stress, signs and symptoms of heat-induce illnesses, first aid procedures, precautionary measures, and other details of the heat management program.

Assessment of program performance and surveillance of heat-induced incidents - In order to identify deficiencies with the heat management program a periodic review is warranted. Input from the workers affected by the program is necessary for the evaluation of the program to be effective. Identification and analysis of the circumstances pertinent to any heat-induce accident or illness is also crucial for correcting program deficiencies.

INDEX

A

abrasive belt manufacturing, 225
abrasive blasting cabinet, ventilation design, 223
accelerometer, 250
adhesives, 159
administrative controls, 21, 178, 236
Adson's maneuvers, 85
age, 18, 29, 50, 57, 68
alcohol use, 57, 69, 162
American Fuel Cell and Coated Fabrics Company, 155
American Podiatric Association, 43
anti-fatigue mats. *See floor mats*
anti-inflammation medications, 19, 23
arm abduction, active or resisted, 85
Armstrong, T.J., 79, 127, 189
aural (ear) temperature measurements. *See ear canal temperature measurements*

B

back strain/sprain, 267
back-pack vacuum cleaners, 247
Belding-Hatch heat stress index (HSI), 162, 278
bi-focal eyeglasses, 53, 57, 62, 69, 74
bicipital tendonitis, 85
Big Bear Grocery Warehouse, 267
biomechanical assessment, 165, 230, 249, 274
black globe temperature (GT), 162
bloating, 167
body map discomfort diagram, 273, 306
Borg scale, 273, 280
breast cancer, 168, 280
buddy system, 317
buffing and polishing wheels, ventilation design, 216
bunion, 45
bursitis, 161

C

calorimetry, 163
carpal tunnel syndrome, 7, 57, 69, 85, 156, 161, 229
case control study, 229
cervical root syndrome, 85
Chaffin, D.B., 43, 274, 291
chair variables, 70, 106
coefficient of friction, 45
colon cancer, 168, 178
computor-generated work orders, 290
computor monitoring. *See electronic performance monitoring*
confined space, 164, 174, 176, 184, 197
confined space classifications, 177, 196
confounders, 15, 68, 72
contact lens, 53, 57, 62
convective heat exchange, 286
conveyors, 20
core body temperature, 163, 278
coronary artery disease, 72
cross sectional study, 67, 229
cumulative trauma disorder (CTD), 20, 156, 161, 228
cycle time, 24, 117, 169

D

deltoid palpatation, 85
department stores, 43
deQuervain's disease, 85, 161
diabetes mellitus, 57, 69
dip tank, ventilation design, 219
disk disease in lower back, 57, 69
disk disease in the neck, 57, 69
dizziness, 167
Dow Jones & Company, 39
dry bulb temperature (DB), 162, 285
dynamic trunk motion analysis, 282

E

ear canal temperature measurement, 164
effective temperature (ET), 162, 278
elbow flexion measurement, 105
electronic performance monitoring, 53, 58, 63, 72, 96
employee modification recommendations, 135
employee turnover rates, 3
energy expenditure modeling, 276
engineering controls, 20, 124, 179, 235
environmental tobacco smoke (ETS), 186. *See also smoking*
epicondyle palpatation, 85
epicondylitis, 7, 85
epidemiologic studies, 67, 71, 113, 229
ergonomic committee, 74, 111, 122
ethanol, 161, 194
ethanol, air monitoring results, 166, 207
evaporative heat loss, 286
exercise, 24
eye gaze angle measurement, 105
eyeglasses, 53, 62

F

fatigue, 169
Findelstein's maneuver, 85
floor mats, 180, 182
floor surfaces, 44
flywheel milling, 111
foot rests, 47, 180
free body diagrams, 250, 261

G

ganglion cysts, 52, 85, 161
gender, 16, 18, 29, 50, 52, 57
gloves, padded, 180

gloves, protective, 167, 175, 183
gout, 57, 69
Guyon Tinel's maneuver, 85
guyon tunnel syndrome, 85

H

hammertoe, 45
hand-held tools, 113, 119
Harley-Davidson Incorporated, 111
heart disease, 72, 186
heart rate monitoring, 276
heat acclimatization, 162, 316
heat balance equation, 277
heat cramps, 315
heat exhaustion, 315
heat rash, 315
heat stress, 162, 185, 276, 285, 290
heat stress criteria, 163
heat stress index. *See Belding-Hatch heat stress index*
heat stress management program, 210
heat stress monitoring results, 167, 208
heat stroke, 315
hoist, chain-operated, 171
hyperabduction, 85
hypersensitivity, 160

I

incentive system, 271
incidence rates, 21, 139
ingrown toenail, 45
inner elbow angle measurement, 105

J

job restructuring, 20
job rotation, 19, 21, 74
job task rotation, 57, 70
joint-related, 52, 85

K

keyboard variables, 106
keystrokes, 53, 58, 64, 67, 72
kidney failure, 57, 69
kneeling, prolonged, 174
knives, use of, 10, 13, 20

L

large paint booth, ventilation design, 222
L5S1 intervertebral joint, 249, 260
lifting rate, 283
light/restricted duty, 24, 289
local exhaust ventilation (LEV), 183, 186
longitudinal study, 71
low back injuries, risk factors, 114, 230
lower extremity disorders, 41

lumbar motion measurement, 275
lung cancer
lupus, 57, 69

M

maximum moment, 283
meatpacking, 3
medical surveillance program, 126, 185
MEK. *See methyl ethyl ketone*
metabolic criteria, 300
metabolic heat load, 163, 195, 284, 286
metabolic heat load production measurement, 163, 276. *See also calorimetry*
metal gratings, as flooring, 44
methyl ethyl ketone (MEK), 155, 159, 194
methyl ethyl ketone (MEK), air monitoring results, 166, 175, 199
Mill's maneuver, 85
motorcycle manufacturing, 114
musculoskeletal disorders, upper extremity, 51, 53, 57, 61, 64, 66, 113, 161

N

nausea, 167
newspaper reporters, 71
NIOSH Lifting Equation calculations, 136, 298, 312
NIOSH 1991 Lifting Equation, 281
noise, 171, 281

O

obesity, 114
open surface tanks, ventilation design, 220
optical scanner, 75
organization characteristics, 62
OSHA 200 injury and illness records, 56, 65, 117, 167, 225, 232, 272, 279
overtime work, 62, 70, 74, 169
oxygen consumption measurement, 276

P

Perdue farms, 3
period prevalence rates, 5, 8, 17, 30
personal protective equipment (PPE). *See gloves, protective and respiratory protection*
Phalen's maneuver, 85
pinch grip, 156, 172, 180
point prevalence rates, 6, 9, 17, 30
portable hand grinding, ventilation design, 217
postural measurements, 59, 101
postural variables, 105
poultry processing plant, 3
powered shears, use of, 156, 179
printing, 37
production standards, 53, 64

proximal interphalangeal joint dysfunction (PIP), 7, 85
psychological stress, 72
psychosocial, 58, 63, 67, 70, 75
psychosocial variables, 95

Q

questionaire, 36, 56, 272, 279, 307

R

race, 16, 18, 29, 51, 68
radial tunnel syndrome, 85
radiant heat gain, 286
razor knives, use of, 171, 180
recreational activities, 68
relative humidity (RH), 285
resisted flexion, or extension, or rotation, 85
respiratory protection, 175, 184
rest pauses, 21, 74. *See also work breaks*
retail trade, 44
rheumatoid arthritis, 52, 57, 69
rotator cuff tendonitis, 6, 85

S

Sancap Abrasives Inc., 225
scissors, use of, 10, 13, 20, 156, 169, 172
shoes, 47
shoulder flexion measurement, 105
SIC 2016 Poultry Processing Plants, 3
SIC 2711 Newspapers: Publishing, or Publishing and Printing, 40
SIC 3068 Fabricated Rubber Products, noc, 157
SIC 3291 Abrasives Products, 225
SIC 3751 Motorcycle Manufacturing, 112
SIC 4813 Telecommunications, 54
SIC 5411 Grocery Stores, Warehouse, 269
SIC 7349 Janitorial Service, 247
skin absorption, 160
skin rash, 156, 167, 178
skin temperature measurements, 251, 265
smoking, 114
solvent degreasing tanks, ventilation design, 218
splint, 24
Spurling's maneuver, 85
standing postures, 46, 170
strength design limit (SDL), 234
supermarket, 43
survivor bias, 3, 15
synovitis, 161

T

tendonitis, 85, 161
tendonitis of the wrist or fingers, 7
tenosynovitis, 161
tension neck syndrome, 6, 85

thoracic outlet syndrome, 85
thyroid disorders, 52, 57, 69, 156
time and motion study, 116
Tinel's maneuver, 85
tool handles, 180
trapezius palpation, 85
1,1,1-trichloroethane, 155, 160, 184, 194
1,1,1-trichloroethane, air monitoring results, 166, 175, 204
trigger finger, 85
trunk motion evaluation, 275
typesetter machines, 41

U

Unico, 247
University of Michigan Static Strength Prediction Program, 230, 234, 245, 281, 298
US West Communications, 51

V

VDTs. *See video display terminals*
vibration, 166
vibration from tools, 177, 119, 156, 171, 174, 180
vibration measurements, 250. *See also accelerometer*
video display terminals (VDTs), 42, 55, 62, 64
videotape analysis of jobs, 116, 166, 225, 230
visual testing, 74
vitamins, 19, 23

W

wet bulb globe temperature index (WGBT), 162, 276, 285
wet bulb temperature (WB), 162
women, 8, 28, 32, 43, 52, 68
work breaks, 70, 248
work habits, 67
work methods analysis, 116
work organization, 67, 75, 93, 125
work pace, 70
work practices, 57, 62, 93, 125, 183, 232, 236
work pressure, 71
work-related musculoskeletal disorders (WRMD). *See musculoskeletal disorders, upper extremity*
work schedule rotation, 231, 316
work sequencing, 243, 271
work station design, 73, 101, 181
work station measurements, 59, 105
worker questionaire. *See questionaire*
working metabolic rate. *See metabolic heat load*
wrist extension measurement, 105
wrist ulnar deviation measurement, 105

Y

Yergason's maneuver, 85

About Government Institutes

Government Institutes, Inc. was founded in 1973 to provide continuing education and practical information for your professional development. Specializing in environmental, health and safety concerns, we recognize that you face unique challenges presented by the ever-increasing number of new laws and regulations and the rapid evolution of new technologies, methods and markets.

Our information and continuing education efforts include a Videotape Distribution Service, over 200 courses held nation-wide throughout the year, and over 250 publications, making us the world's largest publisher in these areas.

Government Institutes, Inc.
4 Research Place, Suite 200
Rockville, MD 20850
(301) 921-2355

Other related books published by Government Institutes:

Health Effects Of Toxic Substances — This comprehensive book provides you with an excellent understanding of the toxicology and industrial hygiene of hazardous materials. Chapters cover: Industrial Toxicology - History and Hazards; Exposure and Entry Routes - Pharmacokinetics I; Distribution, Localization, Biotransformation, Elimination; Dose-Effects and Time-Effects Relationships; Classification, Type, and Limits of Exposure; Action of Toxic Substances Pharmacodynamics; Target Organ Effects; Reproductive Toxins, Mutagens, and Carcinogens; Survey of Common Hazardous Agents I, Toxic Substances; Survey of Common Hazardous Agents II, Physical & Biological Hazards; Types of Environmental Health Hazards; Monitoring of Harmful Agents; Exposure Limits and Personal Protective Equipment; Exposure Control Methods; Medical Monitoring, Treatment, and Management; Risk Assessment. *Softcover, Index, approx. 300 pages, Aug. '95, ISBN: 0-86587-471-9* **$39**

Understanding Workers' Compensation: *A Guide for Safety and Health Professionals* — This book explains in simple and direct terms the Workers Compensation System. It provides a basic understanding of injury prevention, types of injuries, and cost containment strategies. This book includes sample forms, checklists for work site evaluations, and an appendix containing material from the most recent U.S. Chamber of Commerce analysis, comparisons of all state and Canadian provincial laws, policies on rehabilitation, statistics on benefits payable, and waiting periods. A directory of state and provincial workers compensation administrators with full contact information is also included. *Softcover, approx. 150 pages, June '95, ISBN: 0-86587-464-6* **$45**

So You're the Safety Director: An Introduction to Loss Control and Safety Management — Author Michael Manning, a safety veteran and sought-after consultant and speaker, has created an introduction to your bottom-line responsibilities, concentrating on your role in evaluating, managing, and controlling your company's losses and handling the OSHA compliance process. Manning's narrative approach and easy-to-follow writing style make it seem like you've hired him to help you start — or upgrade — your safety program, which is exactly what hundreds of companies have done. Let Manning walk you through the in's and out's of establishing and evaluating your company's safety program: comparing your safety program to those of similar companies, establishing safety committees, involving all employees in your safety program, investigating accidents and preventing their recurrence, managing your compensation costs, preparing for and handling OSHA inspections, and using your company's insurance company as a resource. *Softcover, Index, 150 pages, Oct '95, ISBN: 0-86587-481-6* **$45**

Ergonomic Problems in the Workplace: A Guide to Effective Management — The valuable insights you'll gain from this new book will help you develop and implement your own successful ergonomics program. Now your company can reduce injuries — such as Cumulative Trauma Disorders (CTD) — and reduce the number of workers' compensation claims. In addition, case studies help you learn from the successes and failures of other companies. Table of contents includes: developing an ergonomics program; management commitment; case histories; hazard assessment; cumulative trauma disorders; workplace hazards; hazard prevention and controls; back injuries and material handling; tool selection; ergonomic personal protective equipment; implementing an ergonomics program; medical management; VDTs and office ergonomics; heat stress; training; ADA and ergonomics; working with OSHA on ergonomic issues; and sources of information and assistance. *Softcover, 272 pages, Sept '95, ISBN: 0-86587-474-3* **$59**

Call the above number for our current

Publications (cont.)

OSHA Field Inspection Reference Manual — This new revision of inspection guidelines, previously contained in the OSHA Field Operations Manual, is now being used by OSHA inspectors when checking your facility for compliance. Learn where the inspectors will look, what they'll look for, how they'll evaluate your working conditions, and how they'll actually proceed once inside your facility. *Softcover, 144 pages, Jan '95, ISBN: 0-86587-426-3* **$59**

OSHA Technical Manual, 4th Edition — This inspection manual is used nationwide by OSHA Compliance Safety and Health officers (CSHO's) in checking industry compliance. This new **4th Edition** has been revised, reorganized, and broadened in scope to embrace all safety and health disciplines. The new Manual contains 20 chapters in sections on safety hazards; health hazards; sampling, measurement, methods, and instruments; construction operations; health care facilities; ergonomics; personal protective equipment; and safety and health management. In addition, new chapters include Controlling Lead Exposures in the Construction Industry; Ventilation Investigation, Controlling Occupational Exposure to Hazardous Drugs; Laser Hazard; Industrial Robots and Robot System Safety; and Excavations: Hazard Recognition in Trenching and Shoring. *Softcover, 400 pages, Jan '96, ISBN:0-86587-511-1* **$85**

Safety Made Easy: A Checklist Approach to OSHA Compliance Written by Tex Davis, this book provides a new, simpler way of understanding your requirements under the complex maze of OSHA's workplace safety and health regulations. The easy-to-use format and logical organization make this book ideal for those who are just entering the field of safety compliance as well as for experienced safety professionals. *Softcover, 200 pages, May '95, ISBN: 0-86587-463-8* **$45**

NIOSH Case Studies Series — Conducted by NIOSH, these Health Hazard Evaluations are comprehensive case studies which bring the latest case research and technology to bear on work place health problems that are likely to be encountered at a variety of worksites. Edited by Shirley A. Ness

NIOSH Case Studies in Indoor Air Quality — The case studies cover a range of scenarios involving indoor air quality, one of the areas which pose the greatest challenges to professionals seeking to control them. They have been selected based on a number of variables including type of workplace involved, the activities conducted there, and the causative agents determined to be responsible for the problems. *Softcover, 250 pages, Sept '95, ISBN: 0-86587-482-4* **$79**

NIOSH Case Studies in Bioaerosols — These case studies cover a variety of Microbial Agents encountered in workplace environments, often cited as possible causes of indoor air quality problems. *Softcover, approx 320 pages, Dec '95, ISBN:0-86587-485-0* **$79**

NIOSH Case Studies in Lead — The case studies cover a range of industries (including Lead Paint Abatement, Bronze Casting, Battery Manufacturing, Manufacturing and Recycling, Firing Ranges, and Construction), and evaluations of Cleaning Methods and Solvents. *Softcover, 324 pages, Nov '95, ISBN: 0-86587-484-0* **$79**

Educational Programs

■ Our **COURSES** combine the legal, regulatory, technical, and management aspects of today's key environmental, safety and health issues — such as environmental laws and regulations, environmental management, pollution prevention, OSHA and many other topics. We bring together the leading authorities from industry, business and government to shed light on the problems and challenges you face each day. Please call our Education Department at (301) 921-2345 for more information!

■ Our **TRAINING CONSULTING GROUP** can help audit your ES&H training, develop an ES&H training plan, and customize on-site training courses. Our proven and successful ES&H training courses are customized to fit your organizational and industry needs. Your employees learn key environmental concepts and strategies at a convenient location for 30% of the cost to send them to non-customized, off-site courses. Please call our Training Consulting Group at (301) 921-2366 for more information!